Just The

facts101

Textbook Key Facts

e-Study Guide

by cram101

Textbook NOT Included

Just The

facts101

Textbook Key Facts

Textbook Outlines, Highlights, and Practice Quizzes

Radiation Physics for Medical Physicists

by Ervin B. Podgorsak, 2nd Edition

All "Just the Facts101" Material Written or Prepared by Cram101 Publishing

---

Title Page

Cram101 LEARNING SYSTEM

"Just the Facts101" is a Cram101 publication and tool designed to give you all the facts from your textbooks. Visit Cram101.com for the full practice test for each of your chapters for virtually any of your textbooks.

Cram101 has built custom study tools specific to your textbook. We provide all of the factual testable information and unlike traditional study guides, we will never send you back to your textbook for more information.

YOU WILL NEVER HAVE TO HIGHLIGHT A BOOK AGAIN!

Cram101 StudyGuides

All of the information in this StudyGuide is written specifically for your textbook. We include the key terms, places, people, and concepts... the information you can expect on your next exam!

Want to take a practice test?

Throughout each chapter of this StudyGuide you will find links to cram101.com where you can select specific chapters to take a complete test on, or you can subscribe and get practice tests for up to 12 of your textbooks, along with other exclusive cram101.com tools like problem solving labs and reference libraries.

Cram101.com

Only cram101.com gives you the outlines, highlights, and PRACTICE TESTS specific to your textbook. Cram101.com is an online application where you'll discover study tools designed to make the most of your limited study time.

By purchasing this book, you get 50% off the normal subscription free!. Just enter the promotional code **'DK73DW18024'** on the Cram101.com registration screen.

www.Cram101.com

ISBN(s): 9781467274944. PUBE-4.2012810

Learning System

Radiation Physics for Medical Physicists
Ervin B. Podgorsak, 2nd

CONTENTS

CHAPTER OUTLINE: KEY TERMS, PEOPLE, PLACES, CONCEPTS

Chapter 1

Introduction to Modern Physics

______ Alpha particle

______ Avogadro

______ Vacuum permeability

______ Neutron

______ Planck

______ Planck constant

______ Positron

______ Atomic mass

______ Bohr radius

______ Classical electron radius

______ Compton wavelength

______ Fine structure

______ Rydberg constant

______ Wavelength

______ Physical quantity

______ ELEMENTARY

______ Entity

______ Electric charge

______ Baryon

_______________ Gluon

_______________ Graviton

_______________ Hadron

_______________ Lepton

_______________ Meson

_______________ Muon

_______________ Neutrino

_______________ Non-ionizing radiation

_______________ Quark

_______________ Ionizing radiation

_______________ Linear energy transfer

_______________ Nuclear reactor

_______________ Well logging

_______________ Proton beam

_______________ Charged particle

_______________ Crookes tube

_______________ Delta ray

_______________ Pair production

_______________ Photoelectric effect

Chapter 1. Introduction to Modern Physics

Positron emission

Positron emission tomography

Thermionic emission

Beta decay

Beta particle

Electron pair

Internal conversion

Deuterium

Tritium

Bragg peak

Nuclear reaction

Photon

Annihilation

Bremsstrahlung

GAMMA

Synchrotron

Synchrotron radiation

Becquerel

Curie

	Equivalent
	Equivalent dose
	Sievert
	Weighting
	Brachytherapy
	External
	Radiology
	Relative biological effectiveness
	Effectiveness
	Particle beam
	Atomic number
	Mass number
	Nucleon
	Atomic weight
	Elementary charge
	Carbon dioxide
	Isotone
	Nuclear structure
	Nuclide

Chapter 1. Introduction to Modern Physics

______ Radionuclide

______ Binding energy

______ Isomer

______ Nuclear binding energy

______ Power spectral density

______ Enrico Fermi

______ Lise Meitner

______ Nuclear fission

______ Chain reaction

______ Electromagnetic radiation

______ Work function

______ Galilean transformation

______ Inertial frame of reference

______ Lorentz transformation

______ Acceleration

______ Kinetic energy

______ Diffraction

______ Interference

______ Matter wave

______________ Max Born

______________ Quantum mechanics

______________ Wave function

______________ Mechanical wave

______________ Mechanic

______________ Wave equation

______________ Differential operator

______________ Eigenfunction

______________ Born approximation

______________ Law of cosines

______________ Rutherford scattering

______________ Uncertainty principle

______________ X-ray tube

______________ Exoelectron emission

______________ Ionization

______________ Alpha decay

______________ Carbon nanotube

______________ Ampere

______________ Lorentz force

Chapter 1. Introduction to Modern Physics

_______________ Permittivity

_______________ Magnetism

_______________ Poynting vector

_______________ Probability density function

_______________ Standard deviation

_______________ Deviation

_______________ Cumulative distribution function

_______________ Variance

_______________ Distribution function

_______________ Error function

_______________ Collimator

Alpha particle

Alpha particles consist of two protons and two neutrons bound together into a particle identical to a helium nucleus, which is classically produced in the process of alpha decay, but may be produced also in other ways and given the same name. The alpha particle can be written as He^{2+}, $_4$

$2He^{2+}$ or $_4$

2He (as it is possible that the ion gains electrons from the environment; also, electrons are not important in nuclear chemistry).

The nomenclature is not well defined, and thus not all high-velocity helium nuclei are considered by all authors as alpha particles.

Avogadro

Avogadro (software) is a molecular editor designed for cross-platform use in computational chemistry, molecular modeling, bioinformatics, materials science, and related are. It is extensible through a plugin architecture.

Features

- Molecular builder/editor for Windows, Linux, and Mac OS X.
- All source code is available under the GNU GPL.
- Translations into Chinese, French, German, Italian, Russian, and Spanish.
- Supports multi-threaded rendering and computation.
- Plugin architecture for developers, including rendering, interactive tools, commands, and Python scripts.
- OpenBabel import of files, input generation for multiple computational chemistry packages, crystallography, and biomolecules.

.

Vacuum permeability	The physical constant μ_0, commonly called the vacuum permeability, permeability of free space, or magnetic constant is an ideal, (baseline) physical constant, which is the value of magnetic permeability in a classical vacuum. Vacuum permeability is derived from production of a magnetic field by an electric current or by a moving electric charge and in all other formulas for magnetic-field production in a vacuum. In the reference medium of classical vacuum, μ_0 has an exact defined value: $\mu_0 = 4\pi \times 10^{-7}$ V·s/(A·m) ≈ $1.256\,637\,0614... \times 10^{-6}$ H·m^{-1} or N·A^{-2} or T·m/A or Wb/(A·m) in the SI system of units.
Neutron	The neutron is a subatomic hadron particle which has the symbol n or n^0, no net electric charge and a mass slightly larger than that of a proton. With the exception of hydrogen, nuclei of atoms consist of protons and neutrons, which are therefore collectively referred to as nucleons. The number of protons in a nucleus is the atomic number and defines the type of element the atom forms.
Planck	Planck is a space observatory of the European Space Agency (ESA) and designed to observe the anisotropies of the cosmic microwave background (CMB) over the entire sky, at a high sensitivity and angular resolution. Planck was built in the Cannes Mandelieu Space Center by Thales Alenia Space and created as the third Medium-Sized Mission (M3) of the European Space Agency's Horizon 2000 Scientific Programme. The project, initially called COBRAS/SAMBA, is named in honour of the German physicist Max Planck (1858-1947), who won the Nobel Prize for Physics in 1918.
Planck constant	The Planck constant is a physical constant reflecting the sizes of energy quanta in quantum mechanics. It is named after Max Planck, one of the founders of quantum theory, who discovered it in 1900. Classical statistical mechanics requires the existence of h (but does not define its value).

The Planck constant was first described as the proportionality constant between the energy (E) of a photon and the frequency (v) of its associated electromagnetic wave. This relation between the energy and frequency is called the Planck relation:

$$E = h\nu \,.$$

Positron

The positron is the antiparticle or the antimatter counterpart of the electron. The positron has an electric charge of +1e, a spin of 1/2, and has the same mass as an electron. When a low-energy positron collides with a low-energy electron, annihilation occurs, resulting in the production of two or more gamma ray photons .

Atomic mass

The atomic mass is the mass of a specific isotope, most often expressed in unified atomic mass units. The atomic mass is the total mass of protons, neutrons and electrons in a single atom.

The atomic mass is sometimes incorrectly used as a synonym of relative atomic mass, average atomic mass and atomic weight; these differ subtly from the atomic mass.

Bohr radius

The Bohr radius is a physical constant, approximately equal to the most probable distance between the proton and electron in a hydrogen atom in its ground state. It is named after Niels Bohr, due to its role in the Bohr model of an atom. The precise definition of the Bohr radius is:

$$a_0 = \frac{4\pi\varepsilon_0\hbar^2}{m_e e^2} = \frac{\hbar}{m_e\, c\, \alpha}$$

where:

According to 2010 CODATA the Bohr radius has a value of 5.2917721092(17) × 10^{-11} m (i.e., approximately 53 pm or 0.53 angstroms).

Classical electron radius

The classical electron radius, is based on a classical (i.e., non-quantum) relativistic model of the electron. Its value is calculated as

$$r_\mathrm{e} = \frac{1}{4\pi\varepsilon_0}\frac{e^2}{m_\mathrm{e}c^2} = 2.8179402894(58)\times 10^{-15}\mathrm{m}$$

where e and m_e are the electric charge and the mass of the electron, c is the speed of light, and ε_0 is the permittivity of free space.

In cgs units, this becomes more simply

$$r_\mathrm{e} = \frac{e^2}{m_\mathrm{e}c^2} = 2.8179402894(58)\times 10^{-13}\mathrm{cm}$$

with (to three significant digits)

$$e = 4.80\times 10^{-10}\mathrm{esu}, m_e = 9.11\times 10^{-28}\mathrm{g}, c = 3.00\times 10^{10}\mathrm{cm/s}.$$

Using classical electrostatics, the energy required to assemble a sphere of constant charge density, of radius r_e and charge e is

$$E = \frac{3}{5}\frac{1}{4\pi\varepsilon_0}\frac{e^2}{r_\mathrm{e}}.$$

If the charge is on the surface the energy is

$$E = \frac{1}{2}\frac{1}{4\pi\varepsilon_0}\frac{e^2}{r_\mathrm{e}}.$$

Ignoring the factors 3/5 or 1/2, if this is equated to the relativistic energy of the electron ($E = mc^2$) and solved for r_e , the above result is obtained.

Compton wavelength

The Compton wavelength is a quantum mechanical property of a particle. It was introduced by Arthur Compton in his explanation of the scattering of photons by electrons (a process known as Compton scattering). The Compton wavelength of a particle is equivalent to the wavelength of a photon whose energy is the same as the rest-mass energy of the particle.

Fine structure

In atomic physics, the fine structure describes the splitting of the spectral lines of atoms due to first order relativistic corrections.

The gross structure of line spectra is the line spectra predicted by non-relativistic electrons with no spin. For a hydrogenic atom, the gross structure energy levels only depend on the principal quantum number n.

Rydberg constant

The Rydberg constant, symbol R_∞, is a physical constant relating to atomic spectra, in the science of spectroscopy. The constant first arose as an empirical fitting parameter in the Rydberg formula for the hydrogen spectral series, but Niels Bohr later showed that its value could be calculated from more fundamental constants, explaining the relationship via his 'Bohr model'. As of 2010, R_∞ is the most accurately measured fundamental physical constant.

Wavelength

In physics, the wavelength of a sinusoidal wave is the spatial period of the wave--the distance over which the wave's shape repeats. It is usually determined by considering the distance between consecutive corresponding points of the same phase, such as crests, troughs, or zero crossings, and is a characteristic of both traveling waves and standing waves, as well as other spatial wave patterns. Wavelength is commonly designated by the Greek letter lambda (λ).

Physical quantity

A physical quantity is a physical property of a phenomenon, body, or substance, that can be quantified by measurement.

Definition of a physical quantity
Formally, the International Vocabulary of Metrology, 3rd edition (VIM3) defines quantity as:

Hence the value of a physical quantity q is expressed as the product of a numerical value N_q and a unit of measurement u_q;

$$q = N_q \times \mathrm{u_q}$$

Quantity calculus describes how to do maths with quantities.

Examples

If the temperature T of a body is quantified as 300 kelvin (in which T is the quantity symbol, 300 the value, and K is the unit), this is written

Units and dimensions
Units

Most physical quantities include a unit, but not all - some are dimensionless.

ELEMENTARY

In computational complexity theory, the complexity class ELEMENTARY of elementary recursive functions is the union of the classes in the exponential hierarchy.

$$\begin{aligned}\mathrm{ELEMENTARY} &= \mathrm{EXP} \cup 2\mathrm{EXP} \cup 3\mathrm{EXP} \cup \cdots \\ &= \mathrm{DTIME}(2^n) \cup \mathrm{DTIME}(2^{2^n}) \cup \mathrm{DTIME}(2^{2^{2^n}}) \cup \cdots\end{aligned}$$

The name was coined by László Kalmár, in the context of recursive functions and undecidability; most problems in it are far from elementary. Some natural recursive problems lie outside ELEMENTARY, and are thus NONELEMENTARY. Most notably, there are primitive recursive problems which are not in ELEMENTARY. We know

LOWER-ELEMENTARY $\subsetneq$ EXPTIME $\subsetneq$ ELEMENTARY $\subsetneq$ PR

Whereas ELEMENTARY contains bounded applications of exponentiation (for example, $O(2^{2^n})$), PR allows more general hyper operators (for example, tetration) which are not contained in ELEMENTARY.

Definition

The definitions of elementary recursive functions are the same as for primitive recursive functions, except that primitive recursion is replaced by bounded summation and bounded product.

Entity	An entity is something that exists by itself, although it need not be of material existence. In particular, abstractions and legal fictions are usually regarded as entities. In general, there is also no presumption that an entity is animate.
Electric charge	Electric charge is a physical property of matter that causes it to experience a force when near other electrically charged matter. Electric charge comes in two types, called positive and negative. Two positively charged substances, or objects, experience a mutual repulsive force, as do two negatively charged objects.
Baryon	A baryon is a composite particle made up of three quarks (as distinct from mesons, which comprise one quark and one antiquark). Baryons and mesons belong to the hadron family, which are the quark-based particles. The name 'baryon' comes from the Greek word for 'heavy' (βαρ?ς, barys), because, at the time of their naming, most known elementary particles had lower masses than the baryons.
Gluon	Gluons are elementary particles which act as the exchange particles (or gauge bosons) for the strong force between quarks, analogous to the exchange of photons in the electromagnetic force between two charged particles. Since quarks make up the baryons and the mesons, and the strong interaction takes place between baryons and mesons, one could say that the color force is the source of the strong interaction, or that the strong interaction is like a residual color force which extends beyond the baryons, for example when protons and neutrons are bound together in a nucleus.

In technical terms, they are vector gauge bosons that mediate strong interactions of quarks in quantum chromodynamics (QCD).

Graviton

In physics, the graviton is a hypothetical elementary particle that mediates the force of gravitation in the framework of quantum field theory. If it exists, the graviton must be massless (because the gravitational force has unlimited range) and must be a spin 2 boson. This is because the source of gravitation is the stress-energy tensor, a second-rank tensor, compared to electromagnetism, the source of which is the four-current, a first-rank tensor.

Hadron

In particle physics, a hadron is a composite particle made of quarks held together by the strong force (as atoms and molecules are held together by the electromagnetic force). Hadrons are categorized into two families: baryons (made of three quarks) and mesons (made of one quark and one antiquark).

The best-known hadrons are protons and neutrons (both baryons), which are components of atomic nuclei.

Lepton

A lepton is an elementary particle and a fundamental constituent of matter. The best known of all leptons is the electron which governs nearly all of chemistry as it is found in atoms and is directly tied to all chemical properties. Two main classes of leptons exist: charged leptons (also known as the electron-like leptons), and neutral leptons (better known as neutrinos).

Meson

In particle physics, mesons ('mi?z?nz or 'm?z?nz) are hadronic subatomic particles composed of one quark and one antiquark, bound together by the strong interaction. Because mesons are composed of sub-particles, they have a physical size, with a radius roughly one femtometre, which is about $^2/_3$ the size of a proton or neutron. All mesons are unstable, with the longest-lived lasting for only a few hundredths of a microsecond.

Muon

The muon (; from the Greek letter mu (μ) used to represent it) is an elementary particle similar to the electron, with a unitary negative electric charge and a spin of 1/2. Together with the electron, the tau, and the three neutrinos, it is classified as a lepton. As is the case with other leptons, the muon is not believed to have any sub-structure at all (i.e., is not thought to be composed of any simpler particles).

Neutrino	A neutrino is an electrically neutral, weakly interacting elementary subatomic particle with half-integer spin. The neutrino is denoted by the Greek letter ν (nu). All evidence suggests that neutrinos have mass but that their mass is tiny even by the standards of subatomic particles.
Non-ionizing radiation	Non-ionizing (or non-ionising) radiation refers to any type of electromagnetic radiation that does not carry enough energy per quantum to ionize atoms or molecules--that is, to completely remove an electron from an atom or molecule. Instead of producing charged ions when passing through matter, the electromagnetic radiation has sufficient energy only for excitation, the movement of an electron to a higher energy state. Nevertheless, different biological effects are observed for different types of non-ionizing radiation.
Quark	A quark is an elementary particle and a fundamental constituent of matter. Quarks combine to form composite particles called hadrons, the most stable of which are protons and neutrons, the components of atomic nuclei. Due to a phenomenon known as color confinement, quarks are never directly observed or found in isolation; they can only be found within baryons or mesons.
Ionizing radiation	Ionizing (or ionising) radiation is radiation composed of particles that individually can liberate an electron from an atom or molecule, producing ions, which are atoms or molecules with a net electric charge. These tend to be especially chemically reactive, and the reactivity produces the high biological damage caused per unit of energy of ionizing radiation. The degree and nature of such ionization depends on the energy and type of the individual particles composing the radiation, and less upon the radiation particle number.
Linear energy transfer	Linear energy transfer is a measure of the energy transferred to material as an ionizing particle travels through it. Typically, this measure is used to quantify the effects of ionizing radiation on biological specimens or electronic devices. Linear energy transfer is closely related to stopping power.

Chapter 1. Introduction to Modern Physics

Nuclear reactor	A nuclear reactor is a device to initiate and control a sustained nuclear chain reaction. Most commonly they are used for generating electricity and for the propulsion of ships. Usually heat from nuclear fission is passed to a working fluid (water or gas), which runs through turbines that power either ship's propellers or generators.
Well logging	Well logging is the practice of making a detailed record (a well log) of the geologic formations penetrated by a borehole. The log may be based either on visual inspection of samples brought to the surface (geological logs) or on physical measurements made by instruments lowered into the hole (geophysical logs). Well logging is done during all phases of a wells development; drilling, completing, producing and abandoning.
Proton beam	Proton beams, a type of ion beams, are the result of proton particle acceleration by means of a cyclotron or a synchrotron and can be used e.g. in Proton Beam Therapy for cancer treatment or for proton beam writing in lithography.
Charged particle	In physics, a charged particle is a particle with an electric charge. It may be either a subatomic particle or an ion. A collection of charged particles, or even a gas containing a proportion of charged particles, is called a plasma, which is called the fourth state of matter because its properties are quite different from solids, liquids and gases (plasma is the most common state of matter in the universe).
Crookes tube	A Crookes tube is an early experimental elerical discharge tube, invented by English physicist William Crookes and others around 1869-1875, in which cathode rays, streams of elerons, were discovered. A development of the Geissler tube, it consists of a partially evacuated glass container of various shapes, with two metal elerodes, one at either end. When a high voltage is applied between the elerodes, cathode rays (elerons) travel in straight lines from the cathode to the anode.
Delta ray	A delta ray is sometimes used to describe any recoil particle caused by secondary ionization. The term was coined by J.J. Thomson. It is entirely unrelated to the family of subatomic particles named delta baryon.

Pair production

Pair production refers to the creation of an elementary particle and its antiparticle, usually when a photon (or another neutral boson) interacts with a nucleus. For example an electron and its antiparticle, the positron, may be created. This is allowed, provided there is enough energy available to create the pair - at least the total rest mass energy of the two particles - and that the situation allows both energy and momentum to be conserved.

Photoelectric effect

In the photoelectric effect, electrons are emitted from matter (metals and non-metallic solids, liquids or gases) as a consequence of their absorption of energy from electromagnetic radiation of very short wavelength, such as visible or ultraviolet radiation. Electrons emitted in this manner may be referred to as photoelectrons. First observed by Heinrich Hertz in 1887, the phenomenon is also known as the Hertz effect, although the latter term has fallen out of general use.

Positron emission

Positron emission is converted, via the weak force, to a neutron, releasing a positron (the antimatter counterpart of an electron) and a neutrino.

Isotos which undergo this decay and thereby emit positrons include carbon-11, potassium-40, nitrogen-13, oxygen-15, fluorine-18, and iodine-121. As an example, the following equation describes the beta plus decay of carbon-11 to boron-11, emitting a positron and a neutrino:

Inside protons and neutrons, there are fundamental particles called quarks. The two most common tys of quarks are up quarks, which have a charge of $+^{2}{}_{3}$ and down quarks, with a $-^{1}{}_{3}$ charge.

Positron emission tomography

Positron emission tomography is a nuclear medicine imaging technique that produces a three-dimensional image or picture of functional processes in the body. The system detects pairs of gamma rays emitted indirectly by a positron-emitting radionuclide (tracer), which is introduced into the body on a biologically active molecule. Three-dimensional images of tracer concentration within the body are then constructed by computer analysis.

Thermionic emission

Thermionic emission is the heat-induced flow of charge carriers from a surface or over a pontial-energy barrier. This occurs because the thermal energy given to the carrier overcomes the binding pontial, also known as work function of the metal. The charge carriers can be electrons or ions, and in older lirature are sometimes referred to as 'thermions'.

Chapter 1. Introduction to Modern Physics

Beta decay

In nuclear physics, beta decay is a type of radioactive decay in which a beta particle (an electron or a positron) is emitted from an atom. There are two types of beta decay: beta minus and beta plus. In the case of beta decay that produces an electron emission, it is referred to as beta minus, while in the case of a positron emission as beta plus .

Beta particle

Beta particles are high-energy, high-speed electrons or positrons emitted by certain types of radioactive nuclei such as potassium-40. The beta particles emitted are a form of ionizing radiation also known as beta rays. The production of beta particles is termed beta decay. They are designated by the Greek letter beta (β).

Electron pair

In chemistry, an electron pair consists of two electrons that occupy the same orbital but have opposite spins.

Because electrons are fermions, the Pauli exclusion principle forbids these particles from having exactly the same quantum numbers. Therefore the only way to occupy the same orbital, i.e. have the same orbital quantum numbers, is to differ in the spin quantum number.

Internal conversion

Internal conversion is a radioactive decay process where an excited nucleus interacts with an electron in one of the lower atom orbitals, causing the electron to be emitted from the atom. Thus, in an internal conversion process, a high-energy electron is emitted from the radioactive atom, but without beta decay taking place. For this reason, the high-speed electrons from internal conversion are not beta partles (β partles), since the latter come from beta decay.

Deuterium

Deuterium, is one of two stable isotopes of hydrogen. It has a natural abundance in Earth's oceans of about one atom in 6,420 of hydrogen (~156.25 ppm on an atom basis). Deuterium accounts for approximately 0.0156% (or on a mass basis: 0.0312%) of all naturally occurring hydrogen in Earth's oceans, while the most common isotope (hydrogen-1 or protium) accounts for more than 99.98%.

Tritium

Tritium is a radioactive isotope of hydrogen. The nucleus of tritium contains one proton and two neutrons, whereas the nucleus of protium (by far the most abundant hydrogen isotope) contains one proton and no neutrons. Naturally occurring tritium is extremely rare on Earth, where trace amounts are formed by the interaction of the atmosphere with cosmic rays.

Bragg peak	The Bragg peak is a peak on the Bragg curve which plots the energy loss of ionizing radiation during its travel through matter. For protons, α-rays, and other ion rays, the peak occurs immediately before the particles come to rest. This is called Bragg peak, for William Henry Bragg who discovered it in 1903.
Nuclear reaction	In nuclear physics and nuclear chemistry, a nuclear reaction is semantically considered to be the process in which two nuclei, or else a nucleus of an atom and a subatomic particle (such as a proton, neutron, or high energy electron) from outside the atom, collide to produce products different from the initial particles. In principle, a reaction can involve more than three particles colliding, but because the probability of three or more nuclei to meet at the same time at the same place is much less than for two nuclei, such an event is exceptionally rare. Radioactive decays can be considered to be spontaneous nuclear reactions, in as much as there is such a thing as a spontaneous chemical reaction.
Photon	In physics, a photon is an elementary particle, the quantum of light and all other forms of electromagnetic radiation, and the force carrier for the electromagnetic force. The effects of this force are easily observable at both the microscopic and macroscopic level, because the photon has no rest mass; this allows for interactions at long distances. Like all elementary particles, photons are currently best explained by quantum mechanics and exhibit wave-particle duality, exhibiting properties of both waves and particles.
Annihilation	Annihilation is defined as 'total destruction' or 'complete obliteration' of an object; having its root in the Latin nihil (nothing). A literal translation is 'to make into nothing'. In physics, the word is used to denote the process that occurs when a subatomic particle collides with its respective antiparticle.
Bremsstrahlung	Bremsstrahlung is electromagnetic radiation produced by the deceleration of a charged particle when deflected by another charged particle, typically an electron by an atomic nucleus. The moving particle loses kinetic energy, which is converted into a photon because energy is conserved. The term is also used to refer to the process of producing the radiation.

GAMMA

GAMMA experiment is a study of: a) Primary cosmic ray energy spectra and elemental composition (abundances of the elements) at energies 10^{15}-10^{18}eV (so called knee energy region) ; b) Galactic diffuse gamma-ray intensity at energies 10^{14}-10^{15}eV ; c) Extensive Air Showers (EAS) at the mountain level by the ground-based EAS array and underground muon scintillation counters; d) Hard jets production at energies ~10^{16}eV by the muon multi-core shower events.

GAMMA experiment is deployed on the South side of Mount Aragats in Armenia (Cosmic-ray observatory). The facility consists of a ground-based extensive air shower (EAS) array of 33 surface detection stations and 150 underground muon detectors.

Synchrotron

A synchrotron is a particular type of cyclic particle accelerator originating from the cyclotron in which the guiding magnetic field (bending the particles into a closed path) is time-dependent, being synchronized to a particle beam of increasing kinetic energy. The synchrotron is one of the first accelerator concepts that enable the construction of large-scale facilities, since bending, beam focusing and acceleration can be separated into different components.

The first electron synchrotron was constructed by Edwin McMillan in 1945, although the principle had already been published (unknown to him) in a Russian journal by Vladimir Veksler.

Synchrotron radiation

The electromagnetic radiation emitted when charged particles are accelerated radially () is called synchrotron radiation. It is produced in synchrotrons using bending magnets, undulators and/or wigglers. It is similar to cyclotron radiation except that cyclotron radiation is generated by the acceleration of ultrarelativistic charged particles through magnetic fields.

Becquerel

The becquerel is the SI-derived unit of radioactivity. One Bq is defined as the activity of a quantity of radioactive material in which one nucleus decays per second. The Bq unit is therefore equivalent to s^{-1}. The becquerel is named for Henri Becquerel, who shared a Nobel Prize with Pierre and Marie Curie for their work in discovering radioactivity.

Curie	The curie is a non-SI unit of radioactivity. It is defined as 1 Ci = 3.7×10^{10} decays per second. Its continued use is discouraged. One curie is roughly the activity of 1 gram of the radium isotope ^{226}Ra, a substance studied by the Curies.
Equivalent	The equivalent (symbol: eq or Eq), sometimes termed the molar equivalent, is a unit of amount of substance used in chemistry and the biological sciences. The equivalent is formally defined as the amount of a substance which will either: • react with or supply one mole of hydrogen ions (H^+) in an acid-base reaction; or • react with or supply one mole of eltrons in a redox reaction. The mass of one equivalent of a substance is called its equivalent weight. A historical definition, used espially for the chemical elements, describes an equivalent as the amount of a substance that will react with one gram of hydrogen, or with eight grams of oxygen, or with 35.5 grams (1.25 oz) of chlorine, or displaces any of the three.
Equivalent dose	The equivalent absorb radiation dose, usually shorten to equivalent dose, is a comput average measure of the radiation absorb by a fix mass of biological tissue, that attempts to account for the different biological damage potential of different types of ionizing radiation.

Sievert	The sievert is the International System of Units (SI) SI derived unit of dose equivalent radiation. It attempts to quantitatively evaluate the biological effects of ionizing radiation as opposed to just the absorbed dose of radiation energy, which is measured in gray. It is named after Rolf Maximilian Sievert, a Swedish medical physicist renowned for work on radiation dosage measurement and research into the biological effects of radiation.
Weighting	The process of weighting involves emphasizing the contribution of some aspects of a phenomenon (or of a set of data) to a final effect or result -- giving them 'more weight' in the analysis. That is, rather than each variable in the data contributing equally to the final result, some data are adjusted to contribute more than others. It is analogous to the practice of adding extra weight to one side of a pair of scales to favour a buyer or seller.
Brachytherapy	Brachytherapy, sealed source radiotherapy, curietherapy or endocurietherapy, is a form of radiotherapy where a radiation source is placed inside or next to the area requiring treatment. Brachytherapy is commonly used as an effective treatment for cervical, prostate, breast, and skin cancer and can also be used to treat tumours in many other body sites. Brachytherapy can be used alone or in combination with other therapies such as surgery, External Beam Radiotherapy (EBRT) and chemotherapy.
External	The term external is useful for describing certain algebraic structures. The term comes from the concept of an external binary operation which is a binary operation that draws from some external set. To be more specific, a left external binary operation on S over R is a function $f : R \times S \to S$ and a right external binary operation on S over R is a function $f : S \times R \to S$ where S is the set the operation is defined on, and R is the external set (the set the operation is defined over).
Radiology	Radiology is a medical specialty that employs the use of imaging to both diagnose and treat disease visualised within the human body. Radiologists use an array of imaging technologies (such as x-ray radiography, ultrasound, computed tomography (CT), nuclear medicine, positron emission tomography (PET) and magnetic resonance imaging (MRI) to diagnose or treat diseases. Interventional radiology is the performance of (usually minimally invasive) medical procedures with the guidance of imaging technologies.
Relative biological effectiveness	Relative Biological Effectiveness is a health physics concept introduced in the 1950s, after it was noted that different types of radiation might affect living organisms differently. A higher Relative biological effectiveness equates with greater biological damage for equivalent radiation exposure.

The four primary types of ionizing radiation that were of concern to scientists working in laboratories were photon radiation, beta radiation, neutron radiation, and alpha radiation.

Effectiveness

Effectiveness is the capability of producing a desired result. When something is deemed effective, it means it has an intended or expected outcome, or produces a deep, vivid impression.

Etymology

The origin of the word 'effective' stems from the Latin word effectivus, which means creative, productive or effective.

Particle beam

A particle beam is a stream of charged or neutral particles, in many cases moving at near the speed of light.

There is a difference between creation and control of charged particle beams to neutral particle beams, as only the first type can be manipulated to a sufficient extent by devices based on electromagnetism. The manipulation and diagnostics of charged particle beams at high kinetic energies using particle accelerators are main topics of accelerator physics.

Atomic number

In chemistry d physics, the atomic number is the number of protons found in the nucleus of atom d therefore identical to the charge number of the nucleus. It is conventionally represented by the symbol Z. The atomic number uniquely identifies a chemical element. In atom of neutral charge, the atomic number is also equal to the number of electrons.

Mass number

The mass number also called atomic mass number, is the total number of protons and neutrons (together known as nucleons) in an atomic nucleus. Because protons and neutrons both are baryons, the mass number A is identical with the baryon number B as of the nucleus as of the whole atom or ion. The mass number is different for each different isotope of a chemical element.

Chapter 1. Introduction to Modern Physics

Nucleon	In chemistry and physics, a nucleon is one of the particles that makes up the atomic nucleus. Each atomic nucleus consists of one or more nucleons, and each atom in turn consists of a cluster of nucleons surrounded by one or more electrons. There are two kinds of nucleon: the neutron and the proton.
Atomic weight	Atomic weight is a dimensionless physical quantity, the ratio of the average mass of atoms of an element (from a given source) to 1/12 of the mass of an atom of carbon-12 (known as the unified atomic mass unit). The term is usually used, without further qualification, to refer to the standard atomic weights published at regular intervals by the International Union of Pure and Applied Chemistry (IUPAC) and which are intended to be applicable to normal laboratory materials.
Elementary charge	The elementary charge, usually denoted as e, is the eltric charge carried by a single proton, or equivalently, the negation (opposite) of the eltric charge carried by a single eltron. This elementary charge is a fundamental physical constant. To avoid confusion over its sign, e is sometimes called the elementary positive charge.
Carbon dioxide	Carbon dioxide is a naturally occurring chemical compound composed of two oxygen atoms covalently bonded to a single carbon atom. It is a gas at standard temperature and pressure and exists in Earth's atmosphere in this state, as a trace gas at a concentration of 0.039% by volume. As part of the carbon cycle known as photosynthesis, plants, algae, and cyanobacteria absorb carbon dioxide, light, and water to produce carbohydrate energy for themselves and oxygen as a waste product.
Isotone	Two nuclides are isotones if they have the same neutron number N, but different proton number Z. For example, boron-12 and carbon-13 nuclei both contain 7 neutrons, and so are isotones. Similarly, S-36, Cl-37, Ar-38, K-39, and Ca-40 nuclei are all isotones of 20 because they all contain 20 neutrons. Despite its similarity to the Greek for 'same stretching', the term was formed by replacing the 'p' in 'isotope' with 'n' for 'neutron'.
Nuclear structure	Understanding the structure of the atomic nucleus is one of the central challenges in nuclear physics The liquid drop model

----

This is one of the first models of nuclear structure, proposed by Carl Friedrich von Weizsäcker in 1935. It describes the nucleus as a classical fluid made up of neutro and proto, with an internal repulsive electric force proportional to the number of proto.

Nuclide

A nuclide is an atomic species characterized by the specific constitution of its nucleus, i.e., by its number of protons Z, its number of neutrons N, and its nuclear energy state.

The word nuclide was proposed by Truman P. Kohman in 1947. Doctor Kohman originally suggested nuclide as referring to a 'species of nucleus' defined by containing a certain number of neutrons and protons. The word thus was originally intended to focus on the nucleus.

Radionuclide

A radionuclide is an atom with an unstable nucleus, which is a nucleus characterized by excess energy available to be imparted either to a newly created radiation particle within the nucleus or to an atomic electron. The radionuclide, in this process, undergoes radioactive decay, and emits gamma ray(s) and/or subatomic particles. These particles constitute ionizing radiation.

Binding energy

Binding energy is the mechanical energy required to disassemble a whole into separate parts. A bound system typically has a lower potential energy than its constituent parts; this is what keeps the system together--often this means that energy is released upon the creation of a bound state. The usual convention is that this corresponds to a positive binding energy.

Isomer

In chemistry, isomers are compounds with the same molecular formula but different structural formulas. Isomers do not necessarily share similar properties, unless they also have the same functional groups. There are many different classes of isomers, like stereoisomers, enantiomers, geometrical isomers, etc.

Nuclear binding energy

Nuclear binding energy is the energy required to split a nucleus of an atom into its component parts. The component parts are neutrons and protons, which are collectively called nucleons. The binding energy of nuclei is always a positive number, since all nuclei require net energy to separate them into individual protons and neutrons.

Power spectral density	In statistical signal processing and physics, the spectral density, power spectral density, is a positive real function of a frequency variable associated with a stationary stochastic process, or a deterministic function of time, which has dimensions of power per Hz, or energy per Hz. It is often called simply the spectrum of the signal. Intuitively, the spectral density captures the frequency content of a stochastic process and helps identify periodicities.
Enrico Fermi	Enrico Fermi was an Italian-born, naturalized American physicist particularly known for his work on the development of the first nuclear reactor, Chicago Pile-1, and for his contributions to the development of quantum theory, nuclear and particle physics, and statistical mechanics. He was awarded the 1938 Nobel Prize in Physics for his work on induced radioactivity. Fermi is widely regarded as one of the leading scientists of the 20th century, highly accomplished in both theory and experiment.
Lise Meitner	Lise Meitner, FRS (7 November 1878 - 27 October 1968) was an Austrian, later Swedish, physicist who worked on radioactivity and nuclear physics. Meitner was part of the team that discovered nuclear fission, an achievement for which her colleague Otto Hahn was awarded the Nobel Prize. Meitner is often mentioned as one of the most glaring examples of women's scientific achievement overlooked by the Nobel committee.
Nuclear fission	In nuclear physics and nuclear chemistry, nuclear fission refers to either a nuclear reaction or a radioactive decay process in which the nucleus of an atom splits into smaller parts (lighter nuclei), often producing free neutrons and photons (in the form of gamma rays), and releasing a very large amount of energy, even by the energetic standards of radioactive decay. The two nuclei produced are most often of comparable but slightly different sizes, typically with a mass ratio of products of about 3 to 2, for common fissile isotopes. Most fissions are binary fissions (producing two charged fragments), but occasionally (2 to 4 times per 1000 events), three positively charged fragments are produced, in a ternary fission.
Chain reaction	A chain reaction is a sequence of reactions where a reactive product or by-product causes additional reactions to take place. In a chain reaction, positive feedback leads to a self-amplifying chain of events.

Chain reactions are one way in which systems which are in thermodynamic non-equilibrium can release energy or inease entropy in order to reach a state of higher entropy.

Electromagnetic radiation | Electromagnetic radiation is a form of engy emitted and absorbed by charged particles, which exhibits wave-like behavior as it travels through space. EMR has both electric and magnetic field components, which stand in a fixed ratio of intensity to each oth, and which oscillate in phase ppendicular to each oth and ppendicular to the direction of engy and wave propagation. In vacuum, electromagnetic radiation propagates at a charactistic speed, the speed of light.

Work function | In solid-state physics, the work function is the minimum energy (usually measured in electron volts) needed to remove an electron from a solid to a point immediately outside the solid surface (or energy needed to move an electron from the Fermi level into vacuum). Here 'immediately' means that the final electron position is far from the surface on the atomic scale but still close to the solid on the macroscopic scale. The work function is a characteristic property for any solid phase of a substance with a conduction band (whether empty or partly filled).

Galilean transformation | The Galilean transformation is used to transform between the coordinates of two reference frames which differ only by constant relative motion within the constructs of Newtonian physics. This is the passive transformation point of view. The equations below, although apparently obvious, break down at speeds that approach the speed of light owing to physics described by relativity theory.

Inertial frame of reference | In physics, an inertial frame of reference is a frame of reference that describes time homogeneously and space homogeneously, isotropically, and in a time-independent manner.

All inertial frames are in a state of constant, rectilinear motion with respect to one another; they are not accelerating in the sense that an accelerometer at rest in one would detect zero acceleration. Measurements in one inertial frame can be converted to measurements in another by a simple transformation (the Galilean transformation in Newtonian physics and the Lorentz transformation in special relativity).

Chapter 1. Introduction to Modern Physics

Lorentz transformation	In physics, the Lorentz transformation, according to the theory of special relativity, different measurements of space and time by two observers can be converted into the measurements observed in either frame of reference. It reflects the surprising fact that observers moving at different velocities may measure different distances, elapsed times, and even different orderings of events. The Lorentz transformation was originally the resu of attempts by Lorentz and others to explain how the speed of light was observed to be independent of the reference frame, and to understand the symmetries of the laws of electromagnetism.
Acceleration	In physics, acceleration is the rate of change of velocity with time. In one dimension, acceleration is the rate at which something speeds up or slows down. However, since velocity is a vector, acceleration describes the rate of change of both the magnitude and the direction of velocity.
Kinetic energy	The kinetic energy of an object is the energy which it possesses due to its motion. It is defined as the work needed to accelerate a body of a given mass from rest to its stated velocity. Having gained this energy during its acceleration, the body maintains this kinetic energy unless its speed changes.
Diffraction	Diffraction refers to various phenomena which occur when a wave encounters an obstacle. Italian scientist Francesco Maria Grimaldi coined the word 'diffraction' and was the first to record accurate observations of the phenomenon in 1665. In classical physics, the diffraction phenomenon is described as the apparent bending of waves around small obstacles and the spreading out of waves past small openings. Similar effects occur when light waves travel through a medium with a varying refractive index or a sound wave through one with varying acoustic impedance.

Interference	In communations and electrons, especially in telecommunations, interference is anything whh alters, modifies, or disrupts a signal as it travels along a channel between a source and a receiver. The term typally refers to the addition of unwanted signals to a useful signal. Common examples are: • Electromagnet interference (EMI) • Co-channel interference (CCI), also known as crosstalk • Adjacent-channel interference (ACI) • Intersymbol interference (ISI) • Inter-carrier interference (I), caused by doppler shift in OFDM modulation (multitone modulation). • Common-mode interference (CMI) • Conducted interference Interference is typally but not always distinguished from noise, for example white thermal noise.
Matter wave	In quantum mechanics, the concept of matter waves or de Broglie waves ( ) reflects the wave-particle duality of matter. The theory was proposed by Louis de Broglie in 1924 in his PhD thesis. The de Broglie relations show that the wavelength is inversely proportional to the momentum of a particle and is also called de Broglie wavelength.
Max Born	Max Born was a German-British physicist and mathematician who was instrumental in the development of quantum mechanics. He also made contributions to solid-state physics and optics and supervised the work of a nuer of notable physicists in the 1920s and 30s. Born won the 1954 Nobel Prize in Physics (shared with Walther Bothe).
Quantum mechanics	Quantum mechanics, is a branch of physics dealing with physical phenomena where the action is of the order of Planck constant; quantum mechanics departs from classical mechanics primarily at the atomic and subatomic scales, the so-called quantum realm. It provides a mathematical description of much of the dual particle-like and wave-like behavior and interactions of energy and matter. In advanced topics of quantum mechanics, some of these behaviors are macroscopic and only emerge at very low or very high energies or temperatures.
Wave function	A wave function is a probability amplitude in quantum mechanics describing the quantum state of a particle and how it behaves. Typically, its values are complex numbers and, for a single particle, it is a function of space and time. The laws of quantum mechanics (the Schrödinger equation) describe how the wave function evolves over time.

Chapter 1. Introduction to Modern Physics

Mechanical wave

A mechanical or material wave is a wave that needs a medium to travel. The oscillating material does not move far from its initial equilibrium position, as only the energy is transferred by connected particles. Ocean waves and sound are examples of this phenomenon.

A mechanical wave requires an initial energy input.

Mechanic

A mechanic is a craftsman or technician who uses tools to build or repair machinery. Many mechanics are specialized in a particular field such as auto mechanics, bicycle mechanics, motorcycle mechanics, boiler mechanics, general mechanics, industrial maintenance mechanics (millwrights), air conditioning and refrigeration mechanics, aircraft mechanics, diesel mechanics, and tank mechanics in the armed services. Auto mechanics, for example, have many trades within.

Wave equation

The wave equation is an important second-order linear partial differential equation for the description of waves - as they occur in physics - such as sound waves, light waves and water waves. It arises in fields like acoustics, electromagnetics, and fluid dynamics. Historically, the problem of a vibrating string such as that of a musical instrument was studied by Jean le Rond d'Alembert, Leonhard Euler, Daniel Bernoulli, and Joseph-Louis Lagrange.

Differential operator

In mathematics, a differential operator is an operator defined as a function of the differentiation operator. It is helpful, as a matter of notation first, to consider differentiation as an abstract operation, accepting a function and returning another (in the style of a higher-order function in computer science).

Notations

The most common differential operator is the action of taking the derivative itself.

Eigenfunction

In mathematics, an eigenfunction of a linear operator, A, defined on some function space is any non-zero function f in that space that returns from the operator exactly as is, except for a multiplicative scaling factor. More precisely, one has

$$\mathcal{A}f = \lambda f$$

----

for some scalar, λ, the corresponding eigenvalue. The solution of the differential eigenvalue problem also depends on any boundary conditions required of f .

Born approximation

In scattering theory and, in particular in quantum mechanics, the Born approximation consists of taking the incident field in place of the total field as the driving field at each point in the scatterer. Born approximation is named after Max Born, winner of the 1954 Nobel Prize for physics.

It is the perturtion method applied to scattering by an extended body.

Law of cosines

In trigonometry, the law of cosines relates the lengths of the sides of a plane triangle to the cosine of one of its angles. Using notation as in Fig. 1, the law of cosines says

$$c^2 = a^2 + b^2 - 2ab\cos\gamma$$

where γ denotes the angle contained between sides of lengths a and b and opposite the side of length c.

Some schools also describe the notation as follows:

$$c^2 = a^2 + b^2 - 2ab\cos C$$

Where C represents the same as γ and the rest of the parameters are the same.

Rutherford scattering

In physics, Rutherford scattering is a phenomenon that was explained by Ernest Rutherford in 1911, and led to the development of the Rutherford model (planetary model) of the atom, and eventually to the Bohr model. It is now exploited by the materials analytical technique Rutherford backscattering. Rutherford scattering is also sometimes referred to as Coulomb scattering because it relies only upon static electric (Coulomb) forces, and the minimal distance between particles is set only by this potential.

Uncertainty principle

In quantum mechanics, the uncertainty principle is any of a variety of mathematical inequalities asserting a fundamental limit to the precision with which certain pairs of physical properties of a particle, such as position x and momentum p, can be simultaneously known. The more precisely the position of some particle is determined, the less precisely its momentum can be known, and vice versa. The original heuristic argument that such a limit should exist was given by Werner Heisenberg in 1927. A more formal inequality relating the standard deviation of position σ_x and the standard deviation of momentum σ_p was derived by Earle Hesse Kennard later that year (and independently by Hermann Weyl in 1928),

$$\sigma_x \sigma_p \geq \frac{\hbar}{2},$$

where h is the reduced Planck constant.

X-ray tube

An X-ray tube is a vacuum tube that produces X-rays. They are used in X-ray machines. X-rays are part of the electromagnetic spectrum, an ionizing radiation with wavelengths shorter than ultraviolet light.

Exoelectron emission

In atomic physics, exoelectron emission is a weak electron emission, appearing only from pretreated (irradiated, deformed etc) objects. The pretreatment ('excitation') turns the objects into an unequilibrial state. accompanies the relaxation of these unequilibria.

Ionization

Ionization is the process of converting an atom or molecule into an ion by adding or removing charged particles such as electrons or other ions. A substance may dissociate without necessarily producing ions. As an example, the molecules of table sugar dissociate in water (sugar is dissolved) but exist as intact neutral entities.

Alpha decay

Alpha decay is a type of rioactive decay in which an atomic nucleus emits an alpha particle(two protons and two neutrons) and thereby transforms (or 'decays') into an atom with a mass number 4 less and atomic number 2 less. For example:

238

92U → 234

90Th + 4

2He

which can also be written as:

$^{238}U \rightarrow {}^{234}Th + \alpha$,

since an alpha particle is the same as a helium-4 nucleus, which has mass number 4 and atomic number 2. It also has a charge +2, but the charge is usually not written in nuclear equations, which describe nuclear reactions without considering the electrons. This convention is not meant to imply that the nuclei necessarily occur in neutral atoms.

Carbon nanotube

[E]Experimental observation; [T]Theoretical prediction

The above discussion referred to axial properties of the nanotube, whereas simple geometrical considerations suggest that carbon nanotubes should be much softer in the radial direction than along the tube axis. Indeed, TEM observation of radial elasticity suggested that even the van der Waals forces can deform two adjacent nanotubes. Nanoindentation experiments, performed by several groups on multiwalled carbon nanotubes and tapping/contact mode atomic force microscope measurement performed on single-walled carbon nanotube, indicated Young's modulus of the order of several GPa confirming that Ts are indeed rather soft in the radial direction.

Chapter 1. Introduction to Modern Physics

Ampere
The ampere is the SI unit of electric current (symbol: I) and is one of the seven SI base units. It is named after André-Marie Ampère (1775-1836), French mathematician and physicist, considered the father of electrodynamics. In practice, its name is often shortened to amp.

Lorentz force
In physics, particularly electromagnetism, the Lorentz force is the force on a point charge due to electromagnetic fields.

The first derivation of the Lorentz force is commonly attributed to Oliver Heaviside in 1889, although other historians suggest an earlier origin in an 1865 paper by James Clerk Maxwell. Lorentz derived it a few years after Heaviside.

Permittivity
In electromagnetism, absolute permittivity is the measure of the resistance that is encountered when forming an electric field in a medium. In other words, permittivity is a measure of how an electric field affects, and is affected by, a dielectric medium. The permittivity of a medium describes how much electric field (more correctly, flux) is 'generated' per unit charge in that medium.

Magnetism
Magnetism is a property of materials that respond to an applied magnetic field. Permanent magnets have persistent magnetic fields caused by ferromagnetism. That is the strongest and most familiar type of magnetism.

Poynting vector
In physics, the Poynting vector can be thought of as representing the directional energy flux density (the rate of energy transfer per unit area, in W/m^2) of an electromagnetic field. ts inventor John Henry Poynting. Oliver Heaviside and Nikolay Umov independently co-invented the Poynting vector.

Probability density function
In probability theory, a probability density function or density of a continuous random variable, is a function that describes the relative likelihood for this random variable to take on a given value. The probability for the random variable to fall within a particular region is given by the integral of this variable's density over the region. The probability density function is nonnegative everywhere, and its integral over the entire space is equal to one.

Standard deviation
Standard deviation is a widely used measure of variability or diversity used in statistics and probability theory. It shows how much variation or 'dispersion' exists from the average (mean, or expected value). A low standard deviation indicates that the data points tend to be very close to the mean, whereas high standard deviation indicates that the data points are spread out over a large range of values.

Deviation

In mathematics and statistics, deviation is a measure of difference for interval and ratio variables between the observed value and the mean. The sign of deviation (positive or negative), reports the direction of that difference (it is larger when the sign is positive, and smaller if it is negative). The magnitude of the value indicates the size of the difference.

Cumulative distribution function

In probability theory and statistics, the cumulative distribution function or just distribution function, describes the probability that a real-valued random variable X with a given probability distribution will be found at a value less than or equal to x. Intuitively, it is the 'area so far' function of the probability distribution. Cumulative distribution functions are also used to specify the distribution of multivariate random variables.

Variance

In probability theory and statistics, the variance is a measure of how far a set of numbers is spread out. It is one of several descriptors of a probability distribution, describing how far the numbers lie from the mean (expected value). In particular, the variance is one of the moments of a distribution.

Distribution function

In molecular kinetic theory in physics, a particle's distribution function is a function of seven variables, $f(x, y, z, t; v_x, v_y, v_z)$, which gives the number of particles per unit volume in phase space. It is the number of particles per unit volume having approximately the velocity (v_x, v_y, v_z) near the place (x, y, z) and time (t). The usual normalization of the distribution function is

$$n(x, y, z, t) = \int f \, dv_x \, dv_y \, dv_z$$

$$N(t) = \int n \, dx \, dy \, dz$$

Here, N is the total number of particles and n is the number density of particles - the number of particles per unit volume, or the density divided by the mass of individual particles.

Error function

In mathematics, the error function is a special function (non-elementary) of sigmoid shape which occurs in probability, statistics and partial differential equations. It is dined as:

$$\operatorname{erf}(x) = \frac{2}{\sqrt{\pi}} \int_0^x e^{-t^2} dt.$$

(When x is negative, the integral is interpreted as the negative of the integral from x to zero).

The complementary error function, denoted erfc, is dined as

$$\begin{aligned}\operatorname{erfc}(x) &= 1 - \operatorname{erf}(x) \\ &= \frac{2}{\sqrt{\pi}} \int_x^\infty e^{-t^2} dt.\end{aligned}$$

The imaginary error function, denoted erfi, is dined as

$$\operatorname{erfi}(z) = -i \ \operatorname{erf}(i\, z)$$

The complex error function, denoted w(x) and also known as the Faddeeva function, is dined as

$$w(x) = e^{-x^2} \operatorname{erfc}(-ix) = e^{-x^2}[1 + i \ \operatorname{erfi}(x)]$$

The name 'error function'

The error function is used in measurement theory (using probability and statistics), and although its use in other branches of mathematics has nothing to do with the characterization of measurement errors, the name has stuck.

Collimator

A collimator is a device that narrows a beam of particles or waves. To 'narrow' can mean either to cause the directions of motion to become more aligned in a specific direction (i.e., collimated or parallel) or to cause the spatial cross section of the beam to become smaller.

Optical collimators

In optics, a collimator may consist of a curved mirror or lens with some type of light source and/or an image at its focus.

PRACTICE QUIZ
Chapter 1. Introduction to Modern Physics

---

1. The __________, is based on a classical (i.e., non-quantum) relativistic model of the electron. Its value is calculated as

$$r_e = \frac{1}{4\pi\varepsilon_0}\frac{e^2}{m_e c^2} = 2.8179402894(58) \times 10^{-15}\mathrm{m}$$

where e and m_e are the electric charge and the mass of the electron, c is the speed of light, and ε_0 is the permittivity of free space.

In cgs units, this becomes more simply

$$r_e = \frac{e^2}{m_e c^2} = 2.8179402894(58) \times 10^{-13}\mathrm{cm}$$

with (to three significant digits)

$$e = 4.80 \times 10^{-10}\mathrm{esu}, m_e = 9.11 \times 10^{-28}\mathrm{g}, c = 3.00 \times 10^{10}\mathrm{cm/s}.$$

Using classical electrostatics, the energy required to assemble a sphere of constant charge density, of radius r_e and charge e is

$$E = \frac{3}{5}\frac{1}{4\pi\varepsilon_0}\frac{e^2}{r_e}.$$

If the charge is on the surface the energy is

$$E = \frac{1}{2}\frac{1}{4\pi\varepsilon_0}\frac{e^2}{r_e}.$$

Ignoring the factors 3/5 or 1/2, if this is equated to the relativistic energy of the electron ($E = mc^2$) and solved for r_e , the above result is obtained.

a. Fraunhofer lines
b. Hollow atoms
c. Slater integrals
d. Classical electron radius

2. _________ is a measure of the energy transferred to material as an ionizing particle travels through it. Typically, this measure is used to quantify the effects of ionizing radiation on biological specimens or electronic devices.

_________ is closely related to stopping power.

a. Linear energy transfer
b. Long-lived fission product
c. Michael McKubre
d. Mattauch isobar rule

3. _________ is converted, via the weak force, to a neutron, releasing a positron (the antimatter counterpart of an electron) and a neutrino.

Isotos which undergo this decay and thereby emit positrons include carbon-11, potassium-40, nitrogen-13, oxygen-15, fluorine-18, and iodine-121. As an example, the following equation describes the beta plus decay of carbon-11 to boron-11, emitting a positron and a neutrino:

Inside protons and neutrons, there are fundamental particles called quarks. The two most common tys of quarks are up quarks, which have a charge of $+^2{}_3$ and down quarks, with a $-^1{}_3$ charge.

----

a. Proton emission
b. Radiation burn
c. Positron emission
d. Radioactive Incident Monitoring Network

4. _________ is defined as 'total destruction' or 'complete obliteration' of an object; having its root in the Latin nihil (nothing). A literal translation is 'to make into nothing'.

In physics, the word is used to denote the process that occurs when a subatomic particle collides with its respective antiparticle.

a. Antihydrogen
b. Antiparticle
c. Annihilation
d. Antimatter rocket

5. The _________ is the mass of a specific isotope, most often expressed in unified _________ units. The _________ is the total mass of protons, neutrons and electrons in a single atom.

The _________ is sometimes incorrectly used as a synonym of relative _________, average _________ and atomic weight; these differ subtly from the _________.

a. Atomic mass
b. Tau neutrino
c. Tribimaximal mixing
d. Trimaximal mixing

ANSWER KEY
Chapter 1. Introduction to Modern Physics

---

1. d
2. a
3. c
4. c
5. a

Chapter 2

Coulomb Scattering

- ________ Rutherford scattering
- ________ Rutherford model
- ________ Wavelength
- ________ Bhabha scattering
- ________ Elastic scattering
- ________ Mott scattering
- ________ Spinthariscope
- ________ Hyperbolic trajectory
- ________ Impact parameter
- ________ Angular frequency
- ________ Momentum
- ________ Momentum transfer
- ________ Angular momentum
- ________ Polar
- ________ Fine structure
- ________ Atomic radius
- ________ Statistical model
- ________ Born approximation
- ________ Electrostatic

Chapter 2. Coulomb Scattering

- Uncertainty principle
- Point source
- Avogadro
- Bohr radius
- Spectroscopy
- Binding energy
- Fourier transform
- Charged particle
- Scattering theory
- Electron scattering
- Standard deviation
- Deviation
- Classical electron radius
- Radiation length
- Probability density function
- Niels Bohr

Rutherford scattering

In physics, Rutherford scattering is a phenomenon that was explained by Ernest Rutherford in 1911, and led to the development of the Rutherford model (planetary model) of the atom, and eventually to the Bohr model. It is now exploited by the materials analytical technique Rutherford backscattering. Rutherford scattering is also sometimes referred to as Coulomb scattering because it relies only upon static electric (Coulomb) forces, and the minimal distance between particles is set only by this potential.

Rutherford model

The Rutherford model is a model of the atom devised by Ernest Rutherford. Rutherford directed the famous Geiger-Marsden experiment in 1909, which suggested, upon Rutherford's 1911 analysis, that the so-called 'plum pudding model' of J. J. Thomson of the atom was incorrect. Rutherford's new model for the atom, based on the experimental results, contained the new features of a relatively high central charge concentrated into a very small volume in comparison to the rest of the atom and with this central volume also containing the bulk of the atomic mass of the atom.

Wavelength

In physics, the wavelength of a sinusoidal wave is the spatial period of the wave--the distance over which the wave's shape repeats. It is usually determined by considering the distance between consecutive corresponding points of the same phase, such as crests, troughs, or zero crossings, and is a characteristic of both traveling waves and standing waves, as well as other spatial wave patterns. Wavelength is commonly designated by the Greek letter lambda (λ).

Bhabha scattering

In quantum electrodynamics, Bhabha scattering is the electron-positron scattering process:

$$e^+e^- \rightarrow e^+e^-$$

There are two leading-order Feynman diagrams contributing to this interaction: an annihilation process and a scattering process. The Bhabha scattering rate is used as a luminosity monitor in electron-positron colliders.

Bhabha scattering is named after Indian physicist Homi J. Bhabha.

Elastic scattering

In scattering theory and in particular in particle physics, elastic scattering is one of the specific forms of scattering. In this procs, the kinetic energy of the incident particl is conserved, only their direction of propagation is modified (by interaction with other particl and/or a potential).

Electron elastic scattering

Chapter 2. Coulomb Scattering

When an alpha particle is an incident particle and it is diffracted in the Coulomb potential of atoms and molecul, the elastic scattering procs is called Rutherford scattering.

Mott scattering

Mott scattering, also referred to as spin-coupling inelastic Coulomb scattering, is the separation of the two spin states of an electron beam by scattering the beam off the Coulomb field of heavy ato. It is mostly used to measure the spin polarization of an electron beam.

In lay ter, Mott Scattering is similar to Rutherford Scattering but electrons are used instead of Alpha particles as they have much lower mass (by about 4 orders of magnitude).

Spinthariscope

A Spinthariscope is a device for observing individual nuclear disintegrations caused by the interaction of ionizing radiation with a phosphor or scintillator.

The spinthariscope was invented by William Crookes in 1903. While observing the apparently uniform fluorescence on a zinc sulfide screen created by the radioactive emissions (mostly alpha radiation) of a sample of radium, he spilled some of the radium sample, and, owing to its extreme rarity and cost, he was eager to find and recover it. Upon inspecting the zinc sulfide screen under a microscope, he noticed separate flashes of light created by individual alpha particle collisions with the screen.

Hyperbolic trajectory

In astrodynamics or celestial mechanics a hyperbolic trajectory is a Kepler orbit with the eccentricity greater than 1. Under standard assumptions a body traveling along this trajectory will coast to infinity, arriving there with hyperbolic excess velocity relative to the central body. Similarly to parabolic trajectory all hyperbolic trajectories are also escape trajectories. The specific energy of a hyperbolic trajectory orbit is positive.

Impact parameter

The impact parameter  is defined as the perpendicular distance between the path of a projectile and the center of the field  created by an object that the projectile is approaching . It is often referred to in Nuclear Physics , as well as in Classical Mechanics.

The impact parameter is related to the scattering angle by

where is the velocity of the projectile when it is far from the center, and is its closest distance from the center.

Angular frequency

In physics, angular frequency ω (also referred to by the terms angular speed, radial frequency, circular frequency, orbital frequency, and radian frequency) is a scalar measure of rotation rate. Angular frequency is the magnitude of the vector quantity angular velocity. The term angular frequency vector $\vec{\omega}$ is sometimes used as a synonym for the vector quantity angular velocity.

Momentum

In classical mechanics, linear momentum or translational momentum is the product of the mass and velocity of an object:

$$\mathbf{p} \equiv m\mathbf{v}.$$

Like velocity, linear momentum is a vector quantity, possessing a direction as well as a magnitude. Linear momentum is also a conserved quantity, meaning that if a closed system is not affected by external forces, its total linear momentum cannot change.

Momentum transfer

In particle physics, wave mechanics and optics, momentum transfer is the amount of momentum that one particle gives to another particle.

In the simplest example of scattering of two colliding particles with initial momenta $\vec{p}_{i1}, \vec{p}_{i2}$, resulting in final momenta $\vec{p}_{f1}, \vec{p}_{f2}$, the momentum transfer is given by

$$\vec{q} = \vec{p}_{i1} - \vec{p}_{f1} = \vec{p}_{f2} - \vec{p}_{i2}$$

where the last identity expresses momentum conservation. Momentum transfer is an important quantity because $\Delta x = \hbar/|q|$ is a better measure for the typical distance resolution of the reaction than the momenta themselves.

Angular momentum

In physics, angular momentum, moment of momentum, or rotational momentum is a vector quantity that can be used to describe the overall state of a physical system. The angular momentum L of a particle with respect to some point of origin is

$$\mathbf{L} = \mathbf{r} \times \mathbf{p} = \mathbf{r} \times m\mathbf{v},$$

where r is the particle's position from the origin, p = mv is its linear momentum, and × denotes the cross product.

The angular momentum of a system of particles (e.g. a rigid body) is the sum of angular momenta of the individual particles.

Polar

A Polar (also AM Herculis Star) is a type of cataclysmic variable binary star system with a very strong magnetic field.

	In most cataclysmic variable stars, matter from a main-sequence companion star is gravitationally stripped by a white dwarf star in the form of an accretion disk. In polar systems, the magnetic field of the white dwarf is too strong for an accretion disk to form.
Fine structure	In atomic physics, the fine structure describes the splitting of the spectral lines of atoms due to first order relativistic corrections. The gross structure of line spectra is the line spectra predicted by non-relativistic electrons with no spin. For a hydrogenic atom, the gross structure energy levels only depend on the principal quantum number n.
Atomic radius	The atomic radius of a chemical element is a measure of the size of its atoms, usually the mean or typical distance from the nucleus to the boundy of the surrounding cloud of electrons. Since the boundy is not a well-defined physical entity, there e vious non-equivalent definitions of atomic radius. Depending on the definition, the term may apply only to isolated atoms, or also to atoms in condensed matter, covalently bound in molecules, or in ionized and excited states; and its value may be obtained through experimental measurements, or computed from theoretical models.
Statistical model	A statistical model is a formalization of relationships between variables in the form of mathematical equations. A statistical model describes how one or more random variables are related to one or more random variables. The model is statistical as the variables are not deterministically but stochastically related.
Born approximation	In scattering theory and, in particular in quantum mechanics, the Born approximation consists of taking the incident field in place of the total field as the driving field at each point in the scatterer. Born approximation is named after Max Born, winner of the 1954 Nobel Prize for physics.

It is the perturtion method applied to scattering by an extended body.

Electrostatic

Electrostatics is the branch of physics that deals with the phenomena and properties of stationary or slow-moving (without acceleration) electric charges.

Since classical antiquity, it was known that some materials such as amber attract lightweight particles after rubbing. The Greek word for amber, ?λεκτρον electron, was the source of the word 'electricity'.

Uncertainty principle

In quantum mechanics, the uncertainty principle is any of a variety of mathematical inequalities asserting a fundamental limit to the precision with which certain pairs of physical properties of a particle, such as position x and momentum p, can be simultaneously known. The more precisely the position of some particle is determined, the less precisely its momentum can be known, and vice versa. The original heuristic argument that such a limit should exist was given by Werner Heisenberg in 1927. A more formal inequality relating the standard deviation of position σ_x and the standard deviation of momentum σ_p was derived by Earle Hesse Kennard later that year (and independently by Hermann Weyl in 1928),

$$\sigma_x \sigma_p \geq \frac{\hbar}{2},$$

where h is the reduced Planck constant.

Point source

A point source is a single identifiable localized source of something. A point source has negligible extent, distinguishing it from other source geometries. Sources are called point sources because in mathematical modeling, these sources can usually be approximated as a mathematical point to simplify analysis.

Avogadro

Avogadro (software) is a molecular editor designed for cross-platform use in computational chemistry, molecular modeling, bioinformatics, materials science, and related are. It is extensible through a plugin architecture.

Features

- Molecular builder/editor for Windows, Linux, and Mac OS X.
- All source code is available under the GNU GPL.
- Translations into Chinese, French, German, Italian, Russian, and Spanish.
- Supports multi-threaded rendering and computation.
- Plugin architecture for developers, including rendering, interactive tools, commands, and Python scripts.
- OpenBabel import of files, input generation for multiple computational chemistry packages, crystallography, and biomolecules.

.

Bohr radius

The Bohr radius is a physical constant, approximately equal to the most probable distance between the proton and electron in a hydrogen atom in its ground state. It is named after Niels Bohr, due to its role in the Bohr model of an atom. The precise definition of the Bohr radius is:

$$a_0 = \frac{4\pi\varepsilon_0\hbar^2}{m_e e^2} = \frac{\hbar}{m_e\, c\, \alpha}$$

where:

According to 2010 CODATA the Bohr radius has a value of 5.2917721092(17) × 10^{-11} m (i.e., approximately 53 pm or 0.53 angstroms).

Spectroscopy

Spectroscopy is the study of the interaction between matter and radiated energy. Historically, spectroscopy originated through the study of visible light dispersed according to its wavelength, e.g., by a prism. Later the concept was expanded greatly to comprise any interaction with radiative energy as a function of its wavelength or frequency.

Chapter 2. Coulomb Scattering

Binding energy

Binding energy is the mechanical energy required to disassemble a whole into separate parts. A bound system typically has a lower potential energy than its constituent parts; this is what keeps the system together--often this means that energy is released upon the creation of a bound state. The usual convention is that this corresponds to a positive binding energy.

Fourier transform

The Fourier transform is a mathematical operation with many applications in physics and engineering that expresses a mathematical function of time as a function of frequency, known as its frequency spectrum; Fourier's theorem guarantees that this can always be done. For instance, the transform of a musical chord made up of pure notes (without overtones) expressed as amplitude as a function of time, is a mathematical representation of the amplitudes and phases of the individual notes that make it up. The function of time is oen called the time domain representation, and the frequency spectrum the frequency domain representation.

Charged particle

In physics, a charged particle is a particle with an electric charge. It may be either a subatomic particle or an ion. A collection of charged particles, or even a gas containing a proportion of charged particles, is called a plasma, which is called the fourth state of matter because its properties are quite different from solids, liquids and gases (plasma is the most common state of matter in the universe).

Scattering theory

In mathematics and physics, scattering theory is a framework for udying and underanding the scattering of waves and particles. Prosaically, wave scattering corresponds to the collision and scattering of a wave with some material object, for inance sunlight scattered by rain drops to form a rainbow. Scattering also includes the interaction of billiard balls on a table, the Rutherford scattering (or angle change) of alpha particles by gold nuclei, the Bragg scattering (or diffraction) of electrons and X-rays by a cluer of atoms, and the inelaic scattering of a fission fragment as it traverses a thin foil.

Electron scattering

Electron scattering is the procs whereby an electron is deflected from its original trajectory. As they are charged particl, they are subject to electromagnetic forc.

Phenomena

Electrons can be scattered by other charged particl through the electrostatic Coulomb forc.

Standard deviation	Standard deviation is a widely used measure of variability or diversity used in statistics and probability theory. It shows how much variation or 'dispersion' exists from the average (mean, or expected value). A low standard deviation indicates that the data points tend to be very close to the mean, whereas high standard deviation indicates that the data points are spread out over a large range of values.
Deviation	In mathematics and statistics, deviation is a measure of difference for interval and ratio variables between the observed value and the mean. The sign of deviation (positive or negative), reports the direction of that difference (it is larger when the sign is positive, and smaller if it is negative). The magnitude of the value indicates the size of the difference.
Classical electron radius	The classical electron radius, is based on a classical (i.e., non-quantum) relativistic model of the electron. Its value is calculated as $$r_e = \frac{1}{4\pi\varepsilon_0}\frac{e^2}{m_e c^2} = 2.8179402894(58) \times 10^{-15}\text{m}$$ where e and m_e are the electric charge and the mass of the electron, c is the speed of light, and ε_0 is the permittivity of free space. In cgs units, this becomes more simply $$r_e = \frac{e^2}{m_e c^2} = 2.8179402894(58) \times 10^{-13}\text{cm}$$ with (to three significant digits) $$e = 4.80 \times 10^{-10}\text{esu}, m_e = 9.11 \times 10^{-28}\text{g}, c = 3.00 \times 10^{10}\text{cm/s}.$$

Using classical electrostatics, the energy required to assemble a sphere of constant charge density, of radius r_e and charge e is

$$E = \frac{3}{5}\frac{1}{4\pi\varepsilon_0}\frac{e^2}{r_e}.$$

If the charge is on the surface the energy is

$$E = \frac{1}{2}\frac{1}{4\pi\varepsilon_0}\frac{e^2}{r_e}.$$

Ignoring the factors 3/5 or 1/2, if this is equated to the relativistic energy of the electron ($E = mc^2$) and solved for r_e , the above result is obtained.

Radiation length

In physics, the radiation length is a characteristic of a material, related to the energy loss of high energy, electromagnetic-interacting particles with it.

Definition

High-energy electrons (>~10 MeV) predominantly lose energy in matter by bremsstrahlung, and high-energy photons by e^+e^- pair production. The characteristic amount of matter traversed for these related interactions is called the radiation length X_0, usually measured in $g \cdot cm^{-2}$.

Probability density function

In probability theory, a probability density function or density of a continuous random variable, is a function that describes the relative likelihood for this random variable to take on a given value. The probability for the random variable to fall within a particular region is given by the integral of this variable's density over the region. The probability density function is nonnegative everywhere, and its integral over the entire space is equal to one.

Niels Bohr	Niels Bohr was a Danish physicist who made foundational contributions to understanding atomic structure and quantum mechanics, for which he received the Nobel Prize in Physics in 1922. Bohr mentored and collaborated with many of the top physicists of the century at his institute in Copenhagen. He was part of the British team of physicists working on the Manhattan Project. Bohr married Margrethe Nørlund in 1912, and one of their sons, Aage Bohr, grew up to be an important physicist who in 1975 also received the Nobel Prize.

PRACTICE QUIZ
Chapter 2. Coulomb Scattering

---

1. In physics, _________ is a phenomenon that was explained by Ernest Rutherford in 1911, and led to the development of the Rutherford model (planetary model) of the atom, and eventually to the Bohr model. It is now exploited by the materials analytical technique Rutherford backscattering. _________ is also sometimes referred to as Coulomb scattering because it relies only upon static electric (Coulomb) forces, and the minimal distance between particles is set only by this potential.

   a. Scattering amplitude
   b. Rutherford scattering
   c. Spallation Neutron Source
   d. Spectral energy distribution

2. The _________ is a model of the atom devised by Ernest Rutherford. Rutherford directed the famous Geiger-Marsden experiment in 1909, which suggested, upon Rutherford's 1911 analysis, that the so-called 'plum pudding model' of J. J. Thomson of the atom was incorrect. Rutherford's new model for the atom, based on the experimental results, contained the new features of a relatively high central charge concentrated into a very small volume in comparison to the rest of the atom and with this central volume also containing the bulk of the atomic mass of the atom.

   a. Rutherford scattering
   b. Rutherford model
   c. Shelter Island Conference
   d. Stark effect

3. In physics, _________, moment of momentum, or rotational momentum is a vector quantity that can be used to describe the overall state of a physical system. The _________ L of a particle with respect to some point of origin is

$$\mathbf{L} = \mathbf{r} \times \mathbf{p} = \mathbf{r} \times m\mathbf{v},$$

where r is the particle's position from the origin, p = mv is its linear momentum, and × denotes the cross product.

The _________ of a system of particles (e.g. a rigid body) is the sum of angular momenta of the individual particles.

----

a. Angular momentum coupling
b. Angular momentum
c. Axial symmetry
d. Azimuthal quantum number

4. _________ is the mechanical energy required to disassemble a whole into separate parts. A bound system typically has a lower potential energy than its constituent parts; this is what keeps the system together--often this means that energy is released upon the creation of a bound state. The usual convention is that this corresponds to a positive _________.

a. Branching fraction
b. Nuclear chain reaction
c. Binding energy
d. Chemical shift

5. _________ is the study of the interaction between matter and radiated energy. Historically, _________ originated through the study of visible light dispersed according to its wavelength, e.g., by a prism. Later the concept was expanded greatly to comprise any interaction with radiative energy as a function of its wavelength or frequency.

a. Spectrum analyzer
b. Spectroscopy
c. Static light scattering
d. Thomson scattering

ANSWER KEY
Chapter 2. Coulomb Scattering

---

1. b
2. b
3. b
4. c
5. b

CHAPTER OUTLINE: KEY TERMS, PEOPLE, PLACES, CONCEPTS

Chapter 3

Rutherford-Bohr Model of the Atom

______ Angular momentum

______ Bohr model

______ Planck

______ Planck constant

______ Momentum

______ Principal quantum number

______ Quantum number

______ Bohr radius

______ Fine structure

______ Electronic state

______ Binding energy

______ Rydberg constant

______ Balmer series

______ Lyman series

______ Reduced mass

______ Deuterium

______ Muonium

______ Positronium

______ Tritium

_______ Energy levels

_______ Ground state

_______ Muon

_______ Magnetic quantum number

_______ Spatial quantization

_______ Spin-orbit interaction

_______ Correspondence principle

_______ Stark effect

_______ Zeeman effect

_______ Hartree

_______ Spectroscopic notation

_______ Atomic radius

_______ Effective atomic number

_______ Douglas Hartree

_______ Atomic number

_______ Ionization

_______ Valence electron

_______ X-ray astronomy

_______ Chemical symbol

Chapter 3. Rutherford-Bohr Model of the Atom

______________ Pauli exclusion principle

______________ Actinide

______________ Actinium

______________ Lanthanide

______________ Halogen

______________ Noble gas

______________ Bremsstrahlung

______________ Absorption edge

______________ Thermionic emission

______________ Wave equation

______________ William Duane

Angular momentum

In physics, angular momentum, moment of momentum, or rotational momentum is a vector quantity that can be used to describe the overall state of a physical system. The angular momentum L of a particle with respect to some point of origin is

$$\mathbf{L} = \mathbf{r} \times \mathbf{p} = \mathbf{r} \times m\mathbf{v},$$

where r is the particle's position from the origin, p = mv is its linear momentum, and × denotes the cross product.

The angular momentum of a system of particles (e.g. a rigid body) is the sum of angular momenta of the individual particles.

Bohr model

In atomic physics, the Bohr model, introduced by Niels Bohr in 1913, depicts the atom as a small, positively charged nucleus surrounded by electrons that travel in circular orbits around the nucleus--similar in structure to the solar system, but with electrostatic forces providing attraction, rather than gravity. This was an improvement on the earlier cubic model (1902), the plum-pudding model (1904), the Saturnian model (1904), and the Rutherford model (1911). Since the Bohr model is a quantum-physics-based modification of the Rutherford model, many sources combine the two, referring to the Rutherford-Bohr model.

Planck

Planck is a space observatory of the European Space Agency (ESA) and designed to observe the anisotropies of the cosmic microwave background (CMB) over the entire sky, at a high sensitivity and angular resolution. Planck was built in the Cannes Mandelieu Space Center by Thales Alenia Space and created as the third Medium-Sized Mission (M3) of the European Space Agency's Horizon 2000 Scientific Programme. The project, initially called COBRAS/SAMBA, is named in honour of the German physicist Max Planck (1858-1947), who won the Nobel Prize for Physics in 1918.

Planck constant

The Planck constant is a physical constant reflecting the sizes of energy quanta in quantum mechanics. It is named after Max Planck, one of the founders of quantum theory, who discovered it in 1900. Classical statistical mechanics requires the existence of h (but does not define its value).

The Planck constant was first described as the proportionality constant between the energy (E) of a photon and the frequency (v) of its associated electromagnetic wave. This relation between the energy and frequency is called the Planck relation:

$$E = h\nu\,.$$

Momentum

In classical mechanics, linear momentum or translational momentum is the product of the mass and velocity of an object:

$$\mathbf{p} \equiv m\mathbf{v}.$$

Like velocity, linear momentum is a vector quantity, possessing a direction as well as a magnitude. Linear momentum is also a conserved quantity, meaning that if a closed system is not affected by external forces, its total linear momentum cannot change.

Principal quantum number

In the principal quantum number symbolized as n is the first of a set of quantum numbers (which includes: the principal quantum number, the azimuthal quantum number, the magnetic quantum number, and the spin quantum number) of an atomic orbital. The principal quantum number can only have positive integer values. As n increases, the orbital becomes larger and the electron spends more time farther from the nucleus.

Quantum number

Quantum numbers describe values of conserved quantities in the dynamics of the quantum system. Perhaps the most peculiar aspect of quantum mechanics is the quantization of observable quantities. This is distinguished from classical mechanics where the values can range continuously.

Bohr radius

The Bohr radius is a physical constant, approximately equal to the most probable distance between the proton and electron in a hydrogen atom in its ground state. It is named after Niels Bohr, due to its role in the Bohr model of an atom. The precise definition of the Bohr radius is:

$$a_0 = \frac{4\pi\varepsilon_0\hbar^2}{m_e e^2} = \frac{\hbar}{m_e\,c\,\alpha}$$

----

where:

According to 2010 CODATA the Bohr radius has a value of 5.2917721092(17) × 10^{-11} m (i.e., approximately 53 pm or 0.53 angstroms).

Fine structure

In atomic physics, the fine structure describes the splitting of the spectral lines of atoms due to first order relativistic corrections.

The gross structure of line spectra is the line spectra predicted by non-relativistic electrons with no spin. For a hydrogenic atom, the gross structure energy levels only depend on the principal quantum number n.

Electronic state

Electronic state is a quantum state of a system consisting of electrons (usually orbitals or chemical bonds in crystals or molecul). The state with lowt energy is called ground state, stat with higher energy are excited stat.

Binding energy

Binding energy is the mechanical energy required to disassemble a whole into separate parts. A bound system typically has a lower potential energy than its constituent parts; this is what keeps the system together--often this means that energy is released upon the creation of a bound state. The usual convention is that this corresponds to a positive binding energy.

Rydberg constant

The Rydberg constant, symbol R_∞, is a physical constant relating to atomic spectra, in the science of spectroscopy. The constant first arose as an empirical fitting parameter in the Rydberg formula for the hydrogen spectral series, but Niels Bohr later showed that its value could be calculated from more fundamental constants, explaining the relationship via his 'Bohr model'. As of 2010, R_∞ is the most accurately measured fundamental physical constant.

Balmer series

The Balmer series, is the designation of one of a set of six different named series describing the spectral line emissions of the hydrogen atom. The Balmer series is calculated using the Balmer formula, an empirical equation discovered by Johann Balmer in 1885.

The visible spectrum of light from hydrogen displays four wavelengths, 410 nm, 434'np;nm, 486'np;nm, and 656'np;nm, that correspond to emissions of photons by electrons in excited states transitioning to the quantum level described by the principal quantum number n equals 2. There are also a number of ultraviolet Balmer lines with wavelengths shorter than 400'np;nm.

Lyman series

In physics and chemistry, the Lyman series is the series of transitions and resulting ultraviolet emission lines of the hydrogen atom as an electron goes from $n \geq 2$ to $n = 1$ (where n is the principal quantum number referring to the energy level of the electron). The transitions are named sequentially by Greek letters: from $n = 2$ to $n = 1$ is called Lyman-alpha, 3 to 1 is Lyman-beta, 4 to 1 is Lyman-gamma, etc. ts discoverer, Theodore Lyman.

Reduced mass

In physics, the Reduced mass is the 'effective' inertial mass appearing in the two-body problem of Newtonian mechanics. This is a quantity which allows the two-body problem to be solved as if it were a one-body problem. Note however that the mass deteining the gravitational force is not reduced.

Deuterium

Deuterium, is one of two stable isotopes of hydrogen. It has a natural abundance in Earth's oceans of about one atom in 6,420 of hydrogen (~156.25 ppm on an atom basis). Deuterium accounts for approximately 0.0156% (or on a mass basis: 0.0312%) of all naturally occurring hydrogen in Earth's oceans, while the most common isotope (hydrogen-1 or protium) accounts for more than 99.98%.

Muonium

Muonium is an exotic atom made up of an antimuon and an electron, which was discovered in 1960 and is given the chemical symbol Mu. During the muon's 2.2 μs lifetime, muonium can enter into compounds such as muonium chloride or sodium muonide . Due to the mass difference between the antimuon and the electron, muonium is more similar to atomic hydrogen than positronium .

Positronium

Positronium is a system consisting of an electron and its anti-particle, a positron, bound together into an 'exotic atom'. Being unstable, the two particles annihilate each other to produce two gamma ray photons after an average lifetime of 125 ps or three gamma ray photons after 142 ns in vacuum, depending on the relative spin states of the positron and electron. The orbit of the two particles and the set of energy levels is similar to that of the hydrogen atom (electron and proton).

Tritium	Tritium is a radioactive isotope of hydrogen. The nucleus of tritium contains one proton and two neutrons, whereas the nucleus of protium (by far the most abundant hydrogen isotope) contains one proton and no neutrons. Naturally occurring tritium is extremely rare on Earth, where trace amounts are formed by the interaction of the atmosphere with cosmic rays.
Energy levels	A quantum mechanical system or particle that is bound, confined spatially, can only take on certain discrete values of energy, as opposed to classical particles, which can have any energy. These values are called Energy levels. The term is most commonly used for the Energy levels of electrons in atoms or molecules, which are bound by the electric field of the nucleus.
Ground state	The ground state of a quantum mechanical system is its lowest-energy state; the energy of the ground state is known as the zero-point energy of the system. An excited state is any state with energy greater than the ground state. The ground state of a quantum field theory is usually called the vacuum state or the vacuum.
Muon	The muon (; from the Greek letter mu (μ) used to represent it) is an elementary particle similar to the electron, with a unitary negative electric charge and a spin of 1/2. Together with the electron, the tau, and the three neutrinos, it is classified as a lepton. As is the case with other leptons, the muon is not believed to have any sub-structure at all (i.e., is not thought to be composed of any simpler particles).
Magnetic quantum number	In atomic physics, the magnetic quantum number is the third of a set of quantum numbers (the principal quantum number, the azimuthal quantum number, the magnetic quantum number, and the spin quantum number) which describe the unique quantum state of an electron and is designated by the letter m. The magnetic quantum number denotes the energy levels available within a subshell. Derivation There are a set of quantum numbers associated with the energy states of the atom.
Spatial quantization	In quantum mechanics, spatial quantization is the quantization of angular momentum in three-dimensional space. It results from the fact that the angular momentum of a rigid rotor is expressed in three dimensions, and is quantized.

For a rigid rotor, it is possible to know L^2 (the square of the magnitude of angular momentum) and one of the components (Usually L_z - the z-component of angular momentum) simultaneously because such two quantum mechanical operators commute.

Spin-orbit interaction

In quantum physics, the spin-orbit interaction is any interaction of a particle's spin with its motion. The first and best known example of this is that spin-orbit interaction causes shifts in an electron's atomic energy levels (detectable as a splitting of spectral lines), due to electromagnetic interaction between the electron's spin and the nucleus's electric field, through which it moves. A similar effect, due to the relationship between angular momentum and the strong nuclear force, occurs for protons and neutrons moving inside the nucleus, leading to a shift in their energy levels in the nucleus shell model.

Correspondence principle

In physics, the correspondence principle states that the behavior of systems described by the theory of quantum mechanics (or by the old quantum theory) reproduces classical physics in the limit of large quantum numbers. In other words, it says that for large orbits and for large energies, quantum calculations must agree with classical calculations.

The principle was formulated by Niels Bohr in 1920, though he had previously made use of it as early as 1913 in developing his model of the atom.

Stark effect

The Stark effect is the shifting and splitting of spectral lines of atoms and molecules due to prence of an external static electric field. The amount of splitting and or shifting is called the Stark splitting or Stark shift. In general one distinguishes first- and cond-order Stark effects.

Zeeman effect

The Zeeman effect is the splitting of a spectral line into several components in the presence of a static magnetic field. It is analogous to the Stark effect, the splitting of a spectral line into several components in the presence of an electric field. The Zeeman effect is very important in applications such as nuclear magnetic resonance spectroscopy, electron spin resonance spectroscopy, magnetic resonance imaging (MRI) and Mössbauer spectroscopy.

Hartree

The hartree also known as the Hartree energy, is the atomic unit of energy. It is defined as $2R_\infty hc$, where R_∞ is the Rydberg constant, h is the Planck constant and c is the speed of light. The 2006 CODATA recommended value is E_h = 4.359?743?94 (22) × 10^{-18} J = 27.211?383?86 (68) eV.

Spectroscopic notation	Spectroscopic notation provides various ways to specify atomic ionization states, as well as atomic and molecular orbitals. Ionization states Spectroscopists customarily refer to the spectrum arising from a given ionization state of a given element by the element's symbol followed by a Roman numeral. The numeral I is used for spectral lines associated with the neutral element, II for those from the first ionization state, III for those from the second ionization state, and so on.
Atomic radius	The atomic radius of a chemical element is a measure of the size of its atoms, usually the mean or typical distance from the nucleus to the boundy of the surrounding cloud of electrons. Since the boundy is not a well-defined physical entity, there e vious non-equivalent definitions of atomic radius. Depending on the definition, the term may apply only to isolated atoms, or also to atoms in condensed matter, covalently bound in molecules, or in ionized and excited states; and its value may be obtained through experimental measurements, or computed from theoretical models.
Effective atomic number	Effective atomic number has two different mings: one that is the effective nuclear charge of an atom, and one that calculates the average atomic number for a compound or mixture of materials. Both are abbreviated Z_{eff}. For an atom The effective atomic number Z_{eff}, (sometimes referred to as the effective nuclear charge) of an atom is the number of protons an electron in the element effectively 'sees' due to screening by inner-shell electrons.

Chapter 3. Rutherford-Bohr Model of the Atom

Douglas Hartree

Douglas Hartree PhD, FRS (27 March 1897 - 12 February 1958) was an English mathematician and physicist most famous for the development of numerical analysis and its application to the Hartree-Fock equations of atomic physics and the construction of the meccano differential analyser.

Early life

Douglas Hartree was born in Cambridge, United Kingdom. His father, William, was a lecturer in engineering at Cambridge University and his mother, Eva Rayner, was president of the National Council of Women and mayor of the city of Cambridge.

Atomic number

In chemistry d physics, the atomic number is the number of protons found in the nucleus of atom d therefore identical to the charge number of the nucleus. It is conventionally represented by the symbol Z. The atomic number uniquely identifies a chemical element. In atom of neutral charge, the atomic number is also equal to the number of electrons.

Ionization

Ionization is the process of converting an atom or molecule into an ion by adding or removing charged particles such as electrons or other ions. A substance may dissociate without necessarily producing ions. As an example, the molecules of table sugar dissociate in water (sugar is dissolved) but exist as intact neutral entities.

Valence electron

In chemistry, valence electrons are the electrons of an atom that can participate in the formation of chemical bonds with other atoms. Valence electrons are their 'own' electrons, present in the free neutral atom, that combine with valence electrons of other atoms to form chemical bonds. In a single covalent bond both atoms contribute one valence electron to form a shared pair.

X-ray astronomy

X-ray astronomy is an observational branch of astronomy which deals with the study of X-ray observation and detection from astronomical objects. X-radiation is absorbed by the Earth's atmosphere, so instruments to detect X-rays must be taken to high altitude by balloons, sounding rockets, and satellites. X-ray astronomy is part of space science.

Chemical symbol

A chemical symbol is a 1- or 2-letter internationally agreed code for a chemical element, usually derived from the name of the element, often in Latin.

Only the first letter is capitalised. For example, 'He' is the symbol for helium, 'Pb' for lead, 'W' for tungsten .

Pauli exclusion principle

The Pauli exclusion principle is the quantum mechanical principle that no two identical fermions (particles with half-integer spin) may occupy the same quantum state simultaneously. A more rigorous statement is that the total wave function for two identical fermions is anti-symmetric with respect to exchange of the particles. The principle was formulated by Austrian physicist Wolfgang Pauli in 1925.

Actinide

The actinide, actinium through lawrencium.

The actinide series derives its name from the group 3 element actinium. All but one of the actinides are f-block elements, corresponding to the filling of the 5f electron shell; lawrencium, a d-block element, is also generally considered an actinide.

Actinium

Actinium is a radioactive chemical element with the symbol Ac and atomic number 89, which was discovered in 1899. It was the first non-primordial radioactive element to be isolated. Polonium, radium and radon were observed before actinium, but they were not isolated until 1902. Actinium gave the name to the actinide series, a group of 15 similar elements between actinium and lawrencium in the periodic table.

A soft, silvery-white radioactive metal, actinium reacts rapidly with oxygen and moisture in air forming a white coating of actinium oxide that prevents further oxidation.

Lanthanide

The lanthanide, from lanthanum through lutetium. These fifteen elements, along with the chemically similar elements scandium and yttrium, are often collectively known as the rare earth elements.

The informal chemical symbol Ln is used in general discussions of lanthanide chemistry.

Halogen	The halogens or halogen elements are a series of nonmetal elements from Group 17 IUPAC Style (formerly: VII, VIIA) of the periodic table, comprising fluorine (F), chlorine (Cl), bromine (Br), iodine (I), and astatine (At). The artificially created element 117, provisionally referred to by the systematic name ununseptium, may also be a halogen. The group of halogens is the only periodic table group which contains elements in all three familiar states of matter at standard temperature and pressure.
Noble gas	The noble gases are a group of chemical elements with very similar properties: under standard conditions, they are all odorless, colorless, monatomic gases, with very low chemical reactivity. The six noble gases that occur naturally are helium (He), neon (Ne), argon (Ar), krypton (Kr), xenon (Xe), and the radioactive radon (Rn). For the first six periods of the periodic table, the noble gases are exactly the members of group 18 of the periodic table.
Bremsstrahlung	Bremsstrahlung is electromagnetic radiation produced by the deceleration of a charged particle when deflected by another charged particle, typically an electron by an atomic nucleus. The moving particle loses kinetic energy, which is converted into a photon because energy is conserved. The term is also used to refer to the process of producing the radiation.
Absorption edge	An absorption edge, adsorption discontinuity or adsorption limit is a sharp discontinuity in the absorption spectrum of a substance. These discontinuities occur at wavelengths where the energy of an absorbed photon corresponds to an electronic transition or ionization potential. When the quantum energy of the incident radiation becomes smaller than the work required to eject an electron from one or other quantum states in the constituent absorbing atom the incident radiation ceases to be absorbed by that state.

Thermionic emission	Thermionic emission is the heat-induced flow of charge carriers from a surface or over a pontial-energy barrier. This occurs because the thermal energy given to the carrier overcomes the binding pontial, also known as work function of the metal. The charge carriers can be electrons or ions, and in older lirature are sometimes referred to as 'thermions'.
Wave equation	The wave equation is an important second-order linear partial differential equation for the description of waves - as they occur in physics - such as sound waves, light waves and water waves. It arises in fields like acoustics, electromagnetics, and fluid dynamics. Historically, the problem of a vibrating string such as that of a musical instrument was studied by Jean le Rond d'Alembert, Leonhard Euler, Daniel Bernoulli, and Joseph-Louis Lagrange.
William Duane	William Duane (February 17, 1872, at Philadelphia - March 7, 1935, in Devon, Pennsylvania) was an American physicist. A coworker of Marie Curie, he developed a method for generating quantities of radon in the laboratory. Biography Studies • 1888-1892 University of Pennsylvania • 1892-1895 Harvard University • 1895 Universities of Göttingen (as a Tyndall Fellow) • 1895-1897 Berlin doctor father: Max Planck Academic career • 1898-1907 professor at the University of Colorado • 1908-1913 at the laboratory of Pierre and Marie Curie in Paris • 1913-1917 assistant professor of physicsat Harvard University • 1917-1934 professor of biophysics at Harvard University research activities • radioactivity • X-ray spectroscopy, Duane-Hunt law, relating the minimum wavelength of X-rays to the threshold voltage of the cathode rays that excite them. Death

----

Starting in 1925, Duane began suffering a continual decline in health brought on by diabetes.

----

1. The _________, is the designation of one of a set of six different named series describing the spectral line emissions of the hydrogen atom. The _________ is calculated using the Balmer formula, an empirical equation discovered by Johann Balmer in 1885.

   The visible spectrum of light from hydrogen displays four wavelengths, 410 nm, 434'np;nm, 486'np;nm, and 656'np;nm, that correspond to emissions of photons by electrons in excited states transitioning to the quantum level described by the principal quantum number n equals 2. There are also a number of ultraviolet Balmer lines with wavelengths shorter than 400'np;nm.

   a. Lyman series
   b. Struve-Sahade effect
   c. Balmer series
   d. Chemical shift

2. In quantum physics, the _________ is any interaction of a particle's spin with its motion. The first and best known example of this is that _________ causes shifts in an electron's atomic energy levels (detectable as a splitting of spectral lines), due to electromagnetic interaction between the electron's spin and the nucleus's electric field, through which it moves. A similar effect, due to the relationship between angular momentum and the strong nuclear force, occurs for protons and neutrons moving inside the nucleus, leading to a shift in their energy levels in the nucleus shell model.

   a. Biexciton
   b. Spin-orbit interaction
   c. Spin model
   d. Spin-exchange interaction

3. In atomic physics, the _________, introduced by Niels Bohr in 1913, depicts the atom as a small, positively charged nucleus surrounded by electrons that travel in circular orbits around the nucleus--similar in structure to the solar system, but with electrostatic forces providing attraction, rather than gravity. This was an improvement on the earlier cubic model (1902), the plum-pudding model (1904), the Saturnian model (1904), and the Rutherford model (1911). Since the _________ is a quantum-physics-based modification of the Rutherford model, many sources combine the two, referring to the Rutherford-_________.

   a. Compton scattering
   b. Compton wavelength
   c. Bohr model
   d. Hanle effect

----

4. In physics, _________, moment of momentum, or rotational momentum is a vector quantity that can be used to describe the overall state of a physical system. The _________ L of a particle with respect to some point of origin is

$$\mathbf{L} = \mathbf{r} \times \mathbf{p} = \mathbf{r} \times m\mathbf{v},$$

where r is the particle's position from the origin, p = mv is its linear momentum, and × denotes the cross product.

The _________ of a system of particles (e.g. a rigid body) is the sum of angular momenta of the individual particles.

a. Angular momentum coupling
b. Angular momentum operator
c. Axial symmetry
d. Angular momentum

5. Planck is a space observatory of the European Space Agency (ESA) and designed to observe the anisotropies of the cosmic microwave background (CMB) over the entire sky, at a high sensitivity and angular resolution. Planck was built in the Cannes Mandelieu Space Center by Thales Alenia Space and created as the third Medium-Sized Mission (M3) of the European Space Agency's Horizon 2000 Scientific Programme. The project, initially called COBRAS/SAMBA, is named in honour of the German physicist Max _________(1858-1947), who won the Nobel Prize for Physics in 1918.

a. POLARBEAR
b. Prognoz 9
c. Planck
d. QUaD

ANSWER KEY
Chapter 3. Rutherford-Bohr Model of the Atom

---

1. c
2. b
3. c
4. d
5. c

CHAPTER OUTLINE: KEY TERMS, PEOPLE, PLACES, CONCEPTS
Chapter 4

Production of X Rays

______ Excited state

______ Ground state

______ X-ray astronomy

______ Magnetic dipole

______ Selection rule

______ Dipole

______ Spectroscopy

______ Siegbahn notation

______ Photon

______ Auger effect

______ Energy levels

______ Bremsstrahlung

______ Charged particle

______ Ampere

______ Relativistic electromagnetism

______ Electromagnetism

______ Acceleration

______ Poynting vector

______ Electric field

Chapter 4. Production of X Rays

Radiation intensity

Synchrotron

Synchrotron radiation

Lorentz force

Nuclear reactor

Excited state	Excitation is an elevation in energy level above an arbitrary baseline energy state. In physics there is a specific technical definition for energy level which is often associated with an atom being excited to an excited state. In quantum mechanics an excited state of a system (such as an atom, molecule or nucleus) is any quantum state of the system that has a higher energy than the ground state (that is, more energy than the absolute minimum).
Ground state	The ground state of a quantum mechanical system is its lowest-energy state; the energy of the ground state is known as the zero-point energy of the system. An excited state is any state with energy greater than the ground state. The ground state of a quantum field theory is usually called the vacuum state or the vacuum.
X-ray astronomy	X-ray astronomy is an observational branch of astronomy which deals with the study of X-ray observation and detection from astronomical objects. X-radiation is absorbed by the Earth's atmosphere, so instruments to detect X-rays must be taken to high altitude by balloons, sounding rockets, and satellites. X-ray astronomy is part of space science.
Magnetic dipole	A magnetic dipole is the limit of either a closed loop of electric current or a pair of poles as the dimensions of the source are reduced to zero while keeping the magnetic moment constant. It is a magnetic analogue of the electric dipole, but the analogy is not complete. In particular, a magnetic monopole, the magnetic analogue of an electric charge, has never been observed.
Selection rule	In physics and chemistry a selection rule, formally constrains the possible transitions of a system from one state to another. Selection rules have been derived for electronic, vibrational, and rotational transitions. The selection rules may differ according to the technique used to observe the transition.
Dipole	In physics, there are several kinds of dipoles: • An electric dipole is a separation of positive and negative charges. The simplest example of this is a pair of electric charges of equal magnitude but opposite sign, separated by some (usually small) distance. A permanent electric dipole is called an electret. • A magnetic dipole is a closed circulation of electric current.

Chapter 4. Production of X Rays

Spectroscopy	Spectroscopy is the study of the interaction between matter and radiated energy. Historically, spectroscopy originated through the study of visible light dispersed according to its wavelength, e.g., by a prism. Later the concept was expanded greatly to comprise any interaction with radiative energy as a function of its wavelength or frequency.
Siegbahn notation	The Siegbahn notation is used in X-ray spectroscopy to name the spectral lines that are characteristic to elements. It was created by Manne Siegbahn. The characteristic lines in X-ray emission spectra correspond to atomic electronic transitions where an electron jumps down to an unfilled level in one of the inner shells of an atom.
Photon	In physics, a photon is an elementary particle, the quantum of light and all other forms of electromagnetic radiation, and the force carrier for the electromagnetic force. The effects of this force are easily observable at both the microscopic and macroscopic level, because the photon has no rest mass; this allows for interactions at long distances. Like all elementary particles, photons are currently best explained by quantum mechanics and exhibit wave-particle duality, exhibiting properties of both waves and particles.
Auger effect	The Auger effect is a physical phenomenon in which the transition of an electron in an atom filling in an inner-shell vacancy causes the emission of another electron. When a core electron is removed, leaving a vacancy, an electron from a higher energy level may fall into the vacancy, resulting in a release of energy. Although sometimes this energy is released in the form of an emitted photon, the energy can also be transferred to another electron, which is ejected from the atom.
Energy levels	A quantum mechanical system or particle that is bound, confined spatially, can only take on certain discrete values of energy, as opposed to classical particles, which can have any energy. These values are called Energy levels. The term is most commonly used for the Energy levels of electrons in atoms or molecules, which are bound by the electric field of the nucleus.
Bremsstrahlung	Bremsstrahlung is electromagnetic radiation produced by the deceleration of a charged particle when deflected by another charged particle, typically an electron by an atomic nucleus. The moving particle loses kinetic energy, which is converted into a photon because energy is conserved. The term is also used to refer to the process of producing the radiation.

Charged particle	In physics, a charged particle is a particle with an electric charge. It may be either a subatomic particle or an ion. A collection of charged particles, or even a gas containing a proportion of charged particles, is called a plasma, which is called the fourth state of matter because its properties are quite different from solids, liquids and gases (plasma is the most common state of matter in the universe).
Ampere	The ampere is the SI unit of electric current (symbol: I) and is one of the seven SI base units. It is named after André-Marie Ampère (1775-1836), French mathematician and physicist, considered the father of electrodynamics. In practice, its name is often shortened to amp.
Relativistic electromagnetism	Relativistic electromagnetism is a modern teaching strategy for developing electromagnetic field theory from Coulomb's law and Lorentz transformations. Though Coulomb's law expresses action at a distance, it is an easily understood electric force principle. The more sophisticated view of electromagnetism expressed by electromagnetic fields in spacetime can be approached by applying spacetime symmetries.
Electromagnetism	Electromagnetism is the branch of science concerned with the forces that occur between electrically charged particles. In electromagnetic theory these forces are explained using electromagnetic fields. Electromagnetic force is one of the four fundamental interactions in nature, the other three being the strong interaction, the weak interaction and gravitation.
Acceleration	In physics, acceleration is the rate of change of velocity with time. In one dimension, acceleration is the rate at which something speeds up or slows down. However, since velocity is a vector, acceleration describes the rate of change of both the magnitude and the direction of velocity.
Poynting vector	In physics, the Poynting vector can be thought of as representing the directional energy flux density (the rate of energy transfer per unit area, in W/m^2) of an electromagnetic field. ts inventor John Henry Poynting. Oliver Heaviside and Nikolay Umov independently co-invented the Poynting vector.
Electric field	In physics, an electric field surrounds electrically charged particles and time-varying magnetic fields. The electric field depicts the force exerted on other electrically charged objects by the electrically charged particle the field is surrounding. The concept of an electric field was introduced by Michael Faraday.
Radiation intensity	In electromagnetism, radiation intensity describes the power density that an antenna creates in a particular solid angle. A solid angle is a section of the surface of the imaginary sphere around the antenna.

Given an antenna's power density P, radiation density is calculated by multiplying it with the square of the distance (r^2) from the antenna to the designated solid angle:

$$U = P \cdot r^2$$

This is so because unlike power density, radiation intensity does not depend on distance: because radiation intensity is defined as the power through a solid angle, the decreasing power density over distance (i.e. over r^2 of the imaginary sphere around the antenna) due to the inverse-square law is offset by the increasing area of the solid angle due to the same law.

Synchrotron

A synchrotron is a particular type of cyclic particle accelerator originating from the cyclotron in which the guiding magnetic field (bending the particles into a closed path) is time-dependent, being synchronized to a particle beam of increasing kinetic energy. The synchrotron is one of the first accelerator concepts that enable the construction of large-scale facilities, since bending, beam focusing and acceleration can be separated into different components.

The first electron synchrotron was constructed by Edwin McMillan in 1945, although the principle had already been published (unknown to him) in a Russian journal by Vladimir Veksler.

Synchrotron radiation

The electromagnetic radiation emitted when charged particles are accelerated radially (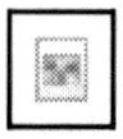) is called synchrotron radiation. It is produced in synchrotrons using bending magnets, undulators and/or wigglers. It is similar to cyclotron radiation except that cyclotron radiation is generated by the acceleration of ultrarelativistic charged particles through magnetic fields.

Lorentz force

In physics, particularly electromagnetism, the Lorentz force is the force on a point charge due to electromagnetic fields.

The first derivation of the Lorentz force is commonly attributed to Oliver Heaviside in 1889, although other historians suggest an earlier origin in an 1865 paper by James Clerk Maxwell. Lorentz derived it a few years after Heaviside.

Nuclear reactor — A nuclear reactor is a device to initiate and control a sustained nuclear chain reaction. Most commonly they are used for generating electricity and for the propulsion of ships. Usually heat from nuclear fission is passed to a working fluid (water or gas), which runs through turbines that power either ship's propellers or generators.

PRACTICE QUIZ
Chapter 4. Production of X Rays

---

1. Excitation is an elevation in energy level above an arbitrary baseline energy state. In physics there is a specific technical definition for energy level which is often associated with an atom being excited to an _________.

   In quantum mechanics an _________ of a system (such as an atom, molecule or nucleus) is any quantum state of the system that has a higher energy than the ground state (that is, more energy than the absolute minimum).

   a. Excited state
   b. Instrumentalist interpretation
   c. Interaction picture
   d. Interatomic Coulombic Decay

2. The _________ of a quantum mechanical system is its lowest-energy state; the energy of the _________ is known as the zero-point energy of the system. An excited state is any state with energy greater than the _________. The _________ of a quantum field theory is usually called the vacuum state or the vacuum.

   a. Ground state
   b. Hamiltonian
   c. Hartman effect
   d. Heisenberg picture

3. In physics, an _________ surrounds electrically charged particles and time-varying magnetic fields. The _________ depicts the force exerted on other electrically charged objects by the electrically charged particle the field is surrounding. The concept of an _________ was introduced by Michael Faraday.

   a. Elementary physics formulae
   b. Electric field
   c. Energy conversion efficiency
   d. Energy current

4. In physics and chemistry a _________, formally constrains the possible transitions of a system from one state to another. _________s have been derived for electronic, vibrational, and rotational transitions. The _________s may differ according to the technique used to observe the transition.

   a. Semi-rigid molecule
   b. Shape resonance
   c. Siascopy
   d. Selection rule

----

5. In physics, a _________ is a particle with an electric charge. It may be either a subatomic particle or an ion. A collection of _________s, or even a gas containing a proportion of _________s, is called a plasma, which is called the fourth state of matter because its properties are quite different from solids, liquids and gases (plasma is the most common state of matter in the universe).

   a. Cooper pair
   b. Non-radiative life time
   c. Charged particle
   d. Peroxocarbonate

ANSWER KEY
Chapter 4. Production of X Rays

1. a
2. a
3. b
4. d
5. c

CHAPTER OUTLINE: KEY TERMS, PEOPLE, PLACES, CONCEPTS
Chapter 5

Two-Particle Collisions

- Neutron
- Charged particle
- Elastic collision
- Inelastic collision
- Nuclear reactor
- Elastic scattering
- Inelastic scattering
- Bremsstrahlung
- Ionization
- Rutherford scattering
- Endothermic
- Exothermic
- Exothermic reaction
- Nuclear binding energy
- Threshold energy
- Binding energy
- Kinetic energy
- Photon
- Fractal

Lichtenberg figure

CHAPTER HIGHLIGHTS: KEY TERMS, PEOPLE, PLACES, CONCEPTS
Chapter 5. Two-Particle Collisions

Neutron	The neutron is a subatomic hadron particle which has the symbol n or n^0, no net electric charge and a mass slightly larger than that of a proton. With the exception of hydrogen, nuclei of atoms consist of protons and neutrons, which are therefore collectively referred to as nucleons. The number of protons in a nucleus is the atomic number and defines the type of element the atom forms.
Charged particle	In physics, a charged particle is a particle with an electric charge. It may be either a subatomic particle or an ion. A collection of charged particles, or even a gas containing a proportion of charged particles, is called a plasma, which is called the fourth state of matter because its properties are quite different from solids, liquids and gases (plasma is the most common state of matter in the universe).
Elastic collision	An elastic collision is an encounter between two bodies in which the total kinetic energy of the two bodies after the encounter is equal to their total kinetic energy before the encounter. Elastic collisions occur only if there is no net conversion of kinetic energy into other forms. During the collision of small objts, kinetic energy is first converted to potential energy associated with a repulsive force between the particles (when the particles move against this force, i.e. the angle between the force and the relative velocity is obtuse), then this potential energy is converted back to kinetic energy (when the particles move with this force, i.e. the angle between the force and the relative velocity is acute).
Inelastic collision	An inelastic collision, in contrast to an elastic collision, is a collision in which kinetic energy is not conserved. In collisions of macroscopic bodies, some kinetic energy is turned into vibrational energy of the atoms, causing a heating effect, and the bodies are deformed. The molecules of a gas or liquid rarely experience perfectly elastic collisions because kinetic energy is exchanged between the molecules' translational motion and their internal degrees of freedom with each collision.

Nuclear reactor	A nuclear reactor is a device to initiate and control a sustained nuclear chain reaction. Most commonly they are used for generating electricity and for the propulsion of ships. Usually heat from nuclear fission is passed to a working fluid (water or gas), which runs through turbines that power either ship's propellers or generators.
Elastic scattering	In scattering theory and in particular in particle physics, elastic scattering is one of the specific forms of scattering. In this procs, the kinetic energy of the incident particl is conserved, only their direction of propagation is modified (by interaction with other particl and/or a potential). Electron elastic scattering When an alpha particle is an incident particle and it is diffracted in the Coulomb potential of atoms and molecul, the elastic scattering procs is called Rutherford scattering.
Inelastic scattering	In particle physics and chemtry, inelastic scattering a fundamental scattering process in which the kinetic energy of an incident particle not conserved (in contrast to elastic scattering). In an inelastic scattering process, some of the energy of the incident particle lost or gained. Although the term htorically related to the concept of inelastic collion in dynamics, the two concepts are quite dtinct; the latter refers to processes in which the total kinetic energy not conserved.
Bremsstrahlung	Bremsstrahlung is electromagnetic radiation produced by the deceleration of a charged particle when deflected by another charged particle, typically an electron by an atomic nucleus. The moving particle loses kinetic energy, which is converted into a photon because energy is conserved. The term is also used to refer to the process of producing the radiation.
Ionization	Ionization is the process of converting an atom or molecule into an ion by adding or removing charged particles such as electrons or other ions. A substance may dissociate without necessarily producing ions. As an example, the molecules of table sugar dissociate in water (sugar is dissolved) but exist as intact neutral entities.
Rutherford scattering	In physics, Rutherford scattering is a phenomenon that was explained by Ernest Rutherford in 1911, and led to the development of the Rutherford model (planetary model) of the atom, and eventually to the Bohr model. It is now exploited by the materials analytical technique Rutherford backscattering. Rutherford scattering is also sometimes referred to as Coulomb scattering because it relies only upon static electric (Coulomb) forces, and the minimal distance between particles is set only by this potential.

Endothermic	In thermodynamics, the word endothermic describes a process or reaction in which the system absorbs energy from the surroundings in the form of heat. It is a modern coinage formed from Greek roots (as is often the case with scientific terminology). The prefix endo- derives from the Greek word 'endon' (?vδov) meaning 'within,' and the latter part of the word comes from the Greek word root 'therm' (θερμ-) meaning 'hot.' Hence it refers to a reaction that needs heat.
Exothermic	In thermodynamics, the term exothermic describes a process or reaction that releases energy from the system, usually in the form of heat, but also in the form of light (e.g. a spark, flame, or explosion), electricity (e.g. a battery), or sound (e.g. burning hydrogen). Its etymology stems from the prefix exo and the Greek word thermasi (meaning 'to heat'). The term exothermic was first coined by Marcellin Berthelot.
Exothermic reaction	An exothermic reaction is a chemical reaction that releases energy in the form of light or heat. It is the opposite of an endothermic reaction. Expressed in a chemical equation: reactants → products + energy Overview An exothermic reaction is a chemical reaction that is accompanied by the release of heat.
Nuclear binding energy	Nuclear binding energy is the energy required to split a nucleus of an atom into its component parts. The component parts are neutrons and protons, which are collectively called nucleons. The binding energy of nuclei is always a positive number, since all nuclei require net energy to separate them into individual protons and neutrons.
Threshold energy	In particle physics, the threshold energy for production of a particle is the minimum kinetic energy a pair of traveling particles must have when they collide. The threshold energy is always grear than or equal to the rest energy of the desired particle. In most cases, since momentum is also conserved, the threshold energy is significantly grear than the rest energy of the desired particle - and thus there will still be considerable kinetic energy in the final particles.
Binding energy	Binding energy is the mechanical energy required to disassemble a whole into separate parts. A bound system typically has a lower potential energy than its constituent parts; this is what keeps the system together--often this means that energy is released upon the creation of a bound state. The usual convention is that this corresponds to a positive binding energy.

Kinetic energy

The kinetic energy of an object is the energy which it possesses due to its motion. It is defined as the work needed to accelerate a body of a given mass from rest to its stated velocity. Having gained this energy during its acceleration, the body maintains this kinetic energy unless its speed changes.

Photon

In physics, a photon is an elementary particle, the quantum of light and all other forms of electromagnetic radiation, and the force carrier for the electromagnetic force. The effects of this force are easily observable at both the microscopic and macroscopic level, because the photon has no rest mass; this allows for interactions at long distances. Like all elementary particles, photons are currently best explained by quantum mechanics and exhibit wave-particle duality, exhibiting properties of both waves and particles.

Fractal

A fractal is a mathematical set that has a fractal dimension that usually exceeds its topological dimension and may fall between the integers. Fractals are typically self-similar patterns, where self-similar means they are 'the same from near as from far' Fractals may be exactly the same at every scale, or as illustrated in Figure 1, they may be nearly the same at different scales. The definition of fractal goes beyond self-similarity per se to exclude trivial self-similarity and include the idea of a detailed pattern repeating itself.[:166; 18]

As mathematical equations, fractals are usually nowhere differentiable, which means that they cannot be measured in traditional ways.

Lichtenberg figure

Lichtenberg figures (Lichtenberg-Figuren, or 'Lichtenberg Dust Figures') are branching electric discharges that sometimes appear on the surface or the interior of insulating materials. he German physicist Georg Christoph Lichtenberg, who originally discovered and studied them. When they were first discovered, it was thought that their characteristic shapes might help to reveal the nature of positive and negative electric 'fluids'.

PRACTICE QUIZ
Chapter 5. Two-Particle Collisions

---

1. An _________ is an encounter between two bodies in which the total kinetic energy of the two bodies after the encounter is equal to their total kinetic energy before the encounter. _________s occur only if there is no net conversion of kinetic energy into other forms.

   During the collision of small objts, kinetic energy is first converted to potential energy associated with a repulsive force between the particles (when the particles move against this force, i.e. the angle between the force and the relative velocity is obtuse), then this potential energy is converted back to kinetic energy (when the particles move with this force, i.e. the angle between the force and the relative velocity is acute).

   a. Elastic collision
   b. Energy conversion efficiency
   c. Energy current
   d. Inertial frame of reference

2. The _________ is a subatomic hadron particle which has the symbol n or n^0, no net electric charge and a mass slightly larger than that of a proton. With the exception of hydrogen, nuclei of atoms consist of protons and _________s, which are therefore collectively referred to as nucleons. The number of protons in a nucleus is the atomic number and defines the type of element the atom forms.

   a. Nucleon
   b. Pentaquark
   c. Neutron
   d. Roper resonance

3. An _________, in contrast to an elastic collision, is a collision in which kinetic energy is not conserved.

   In collisions of macroscopic bodies, some kinetic energy is turned into vibrational energy of the atoms, causing a heating effect, and the bodies are deformed.

   The molecules of a gas or liquid rarely experience perfectly elastic collisions because kinetic energy is exchanged between the molecules' translational motion and their internal degrees of freedom with each collision.

----

a. Inelastic scattering
b. Inflaton
c. Inelastic collision
d. Ionization cooling

4. In physics, a _________ is a particle with an electric charge. It may be either a subatomic particle or an ion. A collection of _________s, or even a gas containing a proportion of _________s, is called a plasma, which is called the fourth state of matter because its properties are quite different from solids, liquids and gases (plasma is the most common state of matter in the universe).

a. Cooper pair
b. Charged particle
c. Non-radiative recombination
d. Peroxocarbonate

ANSWER KEY
Chapter 5. Two-Particle Collisions

---

1. a
2. c
3. c
4. b

CHAPTER OUTLINE: KEY TERMS, PEOPLE, PLACES, CONCEPTS

Chapter 6

Interactions of Charged Particles with Matter

______ Bremsstrahlung

______ Charged particle

______ Momentum

______ Momentum transfer

______ Angular momentum

______ Classical electron radius

______ Kinetic energy

______ Annihilation

______ X-rays

______ Path length

______ Penetration

______ Penetration depth

______ External

______ Elastic scattering

______ X-ray astronomy

______ Inverse

______ Photoelectric effect

______ Filtration

______ Ionization

Ionization chamber

Semiconductor

Semiconductor detector

CHAPTER HIGHLIGHTS: KEY TERMS, PEOPLE, PLACES, CONCEPTS
Chapter 6. Interactions of Charged Particles with Matter

Bremsstrahlung

Bremsstrahlung is electromagnetic radiation produced by the deceleration of a charged particle when deflected by another charged particle, typically an electron by an atomic nucleus. The moving particle loses kinetic energy, which is converted into a photon because energy is conserved. The term is also used to refer to the process of producing the radiation.

Charged particle

In physics, a charged particle is a particle with an electric charge. It may be either a subatomic particle or an ion. A collection of charged particles, or even a gas containing a proportion of charged particles, is called a plasma, which is called the fourth state of matter because its properties are quite different from solids, liquids and gases (plasma is the most common state of matter in the universe).

Momentum

In classical mechanics, linear momentum or translational momentum is the product of the mass and velocity of an object:

$$\mathbf{p} \equiv m\mathbf{v}.$$

Like velocity, linear momentum is a vector quantity, possessing a direction as well as a magnitude. Linear momentum is also a conserved quantity, meaning that if a closed system is not affected by external forces, its total linear momentum cannot change.

Momentum transfer

In particle physics, wave mechanics and optics, momentum transfer is the amount of momentum that one particle gives to another particle.

In the simplest example of scattering of two colliding particles with initial momenta $\vec{p}_{i1}, \vec{p}_{i2}$, resulting in final momenta $\vec{p}_{f1}, \vec{p}_{f2}$, the momentum transfer is given by

$$\vec{q} = \vec{p}_{i1} - \vec{p}_{f1} = \vec{p}_{f2} - \vec{p}_{i2}$$

where the last identity expresses momentum conservation. Momentum transfer is an important quantity because $\Delta x = \hbar/|q|$ is a better measure for the typical distance resolution of the reaction than the momenta themselves.

Angular momentum	In physics, angular momentum, moment of momentum, or rotational momentum is a vector quantity that can be used to describe the overall state of a physical system. The angular momentum L of a particle with respect to some point of origin is $$\mathbf{L} = \mathbf{r} \times \mathbf{p} = \mathbf{r} \times m\mathbf{v} \,,$$ where r is the particle's position from the origin, p = mv is its linear momentum, and × denotes the cross product. The angular momentum of a system of particles (e.g. a rigid body) is the sum of angular momenta of the individual particles.
Classical electron radius	The classical electron radius, is based on a classical (i.e., non-quantum) relativistic model of the electron. Its value is calculated as $$r_e = \frac{1}{4\pi\varepsilon_0}\frac{e^2}{m_e c^2} = 2.8179402894(58) \times 10^{-15}\mathrm{m}$$ where e and m_e are the electric charge and the mass of the electron, c is the speed of light, and ε_0 is the permittivity of free space. In cgs units, this becomes more simply $$r_e = \frac{e^2}{m_e c^2} = 2.8179402894(58) \times 10^{-13}\mathrm{cm}$$ with (to three significant digits) $$e = 4.80 \times 10^{-10}\mathrm{esu}, m_e = 9.11 \times 10^{-28}\mathrm{g}, c = 3.00 \times 10^{10}\mathrm{cm/s}\,.$$

Using classical electrostatics, the energy required to assemble a sphere of constant charge density, of radius r_e and charge e is

$$E = \frac{3}{5} \frac{1}{4\pi\varepsilon_0} \frac{e^2}{r_e}.$$

If the charge is on the surface the energy is

$$E = \frac{1}{2} \frac{1}{4\pi\varepsilon_0} \frac{e^2}{r_e}.$$

Ignoring the factors 3/5 or 1/2, if this is equated to the relativistic energy of the electron ($E = mc^2$) and solved for r_e , the above result is obtained.

Kinetic energy

The kinetic energy of an object is the energy which it possesses due to its motion. It is defined as the work needed to accelerate a body of a given mass from rest to its stated velocity. Having gained this energy during its acceleration, the body maintains this kinetic energy unless its speed changes.

Annihilation

Annihilation is defined as 'total destruction' or 'complete obliteration' of an object; having its root in the Latin nihil (nothing). A literal translation is 'to make into nothing'.

In physics, the word is used to denote the process that occurs when a subatomic particle collides with its respective antiparticle.

X-rays

X-radiation (composed of X-rays) is a form of electromagnetic radiation. X-rays have a wavelength in the range of 10 to 0.01 nanometers, corresponding to frequencies in the range 30 petahertz to 30 exahertz (3×10^{16} Hz to 3×10^{19} Hz) and energies in the range 120 eV to 120 keV. They are shorter in wavelength than UV rays. In many languages, X-radiation is called Röntgen radiation after Wilhelm Conrad Röntgen, who is generally credited as their discoverer, and who had called them X-rays to signify an unknown type of radiation.[:1-2]

X-rays from about 0.12 to 12 keV (10 to 0.10 nm wavelength), are classified as soft X-rays, and from about 12 to 120 keV (0.10 to 0.010 nm wavelength) as hard X-rays, due to their penetrating abilities.

Path length

In chemistry, the path length is defined as the distance that light (UVVIS) travels through a same in an analytical cell. Typically, a same cell is made of quartz, glass, or a astic rhombic cuvette with a volume typically ranging from 0.1 mL to 10 mL or larger used in a spectrophotometer. For the purposes of spectrophotometry (i.e. when making calculations using the Beer-Lambert law) the path length is measured in centimeters (rather than in meters).

Penetration

Strictly speaking penetration occurs when a projectile enters a target without passing through it and perforation occurs when the projectile completely passes through the target, but the word penetration is commonly used to refer to either.

Penetration into a semi-infinite or massive target is penetration (in the strict sense of the word) of targets so thick that the level of penetration is not affected by the target's thickness. There is a transition region between semi-infinite penetration and perforation, in which the target is not perforated but the projectile, as it nears the back face of the target, meets reduced resistance and is capable of penetrating a greater distance than it would in a semi-infinite target.

Penetration depth

Penetration Depth is a measure of how deep light or any electromagnetic radiation can penetrate into a material. It is defined as the depth at which the intensity of the radiation inside the material falls to 1/e (about 37%) of its original value at (or more properly, just beneath) the surface.

When electromagnetic radiation is incident on the surface of a material, it may be (partly) reflected from that surface and there will be a field containing energy transmitted into the material.

External

The term external is useful for describing certain algebraic structures. The term comes from the concept of an external binary operation which is a binary operation that draws from some external set. To be more specific, a left external binary operation on S over R is a function $f : R \times S \rightarrow S$ and a right external binary operation on S over R is a function $f : S \times R \rightarrow S$ where S is the set the operation is defined on, and R is the external set (the set the operation is defined over).

Elastic scattering

In scattering theory and in particular in particle physics, elastic scattering is one of the specific forms of scattering. In this procs, the kinetic energy of the incident particl is conserved, only their direction of propagation is modified (by interaction with other particl and/or a potential).

Electron elastic scattering

When an alpha particle is an incident particle and it is diffracted in the Coulomb potential of atoms and molecul, the elastic scattering procs is called Rutherford scattering.

X-ray astronomy

X-ray astronomy is an observational branch of astronomy which deals with the study of X-ray observation and detection from astronomical objects. X-radiation is absorbed by the Earth's atmosphere, so instruments to detect X-rays must be taken to high altitude by balloons, sounding rockets, and satellites. X-ray astronomy is part of space science.

Inverse

In many contexts in mathematics the term inverse indicates the opposite of something. This word and its derivatives are used widely in mathematics, as illustrated below.

- Inverse element of an element x with respect to a binary operation * with identity element e is an element y such that x * y = y * x = e.

Photoelectric effect	In the photoelectric effect, electrons are emitted from matter (metals and non-metallic solids, liquids or gases) as a consequence of their absorption of energy from electromagnetic radiation of very short wavelength, such as visible or ultraviolet radiation. Electrons emitted in this manner may be referred to as photoelectrons. First observed by Heinrich Hertz in 1887, the phenomenon is also known as the Hertz effect, although the latter term has fallen out of general use.
Filtration	Filtration is commonly the mechanical or physical operation which is used for the separation of solids from fluids (liquids or gases) by interposing a medium through which only the fluid can pass. Oversize solids in the fluid are retained, but the separation is not complete; solids will be contaminated with some fluid and filtrate will contain fine particles (depending on the pore size and filter thickness). Filtration is also used to describe some biological processes, especially in water treatment and sewage treatment in which undesirable constituents are removed by adsorption into a biological film grown on or in the filter medium.
Ionization	Ionization is the process of converting an atom or molecule into an ion by adding or removing charged particles such as electrons or other ions. A substance may dissociate without necessarily producing ions. As an example, the molecules of table sugar dissociate in water (sugar is dissolved) but exist as intact neutral entities.
Ionization chamber	The ionization chamber is the simplest of all gas-filled radiation detectors, and is used for the detection or measurement of ionizing radiation. Conventionally, the term 'ionization chamber' is used exclusively to describe those detectors whh collect ion pairs from gases. Construction and operation An ionization chamber is an instrument constructed to measure charge from the number of ions within a medium (whh we will consider to be gaseous, but can also be solid or liquid).
Semiconductor	A semiconductor is a material with electrical conductivity intermediate in magnitude between that of a conductor and an insulator. This means a conductivity roughly in the range of 10^3 to 10^{-8} siemens per centimeter. Semiconductor materials are the foundation of modern electronics, including radio, computers, telephones, and many other devices.
Semiconductor detector	A semiconductor detector is a device that uses a semiconductor to detect traversing charged particles or the absorption of photons. In the field of particle physics, these detectors are usually known as silicon detectors.

When their sensitive structures are based on a single diode, they are called semiconductor diode detectors.

----

1. _________ is an observational branch of astronomy which deals with the study of X-ray observation and detection from astronomical objects. X-radiation is absorbed by the Earth's atmosphere, so instruments to detect X-rays must be taken to high altitude by balloons, sounding rockets, and satellites. _________ is part of space science.

   a. Cygnus Loop
   b. Cygnus X-3
   c. X-ray astronomy
   d. G117-B15A

2. The term external is useful for describing certain algebraic structures. The term comes from the concept of an external binary operation which is a binary operation that draws from some _________set. To be more specific, a left external binary operation on S over R is a function $f : R \times S \rightarrow S$ and a right external binary operation on S over R is a function $f : S \times R \rightarrow S$ where S is the set the operation is defined on, and R is the _________set (the set the operation is defined over).

   a. Idempotence
   b. If and only if
   c. Intersection
   d. External

3. _________ is electromagnetic radiation produced by the deceleration of a charged particle when deflected by another charged particle, typically an electron by an atomic nucleus. The moving particle loses kinetic energy, which is converted into a photon because energy is conserved. The term is also used to refer to the process of producing the radiation.

   a. Bremsstrahlung
   b. Chaotic scattering
   c. Core-excited shape resonance
   d. Coulomb collision

4. In chemistry, the _________ is defined as the distance that light (UVVIS) travels through a same in an analytical cell. Typically, a same cell is made of quartz, glass, or a astic rhombic cuvette with a volume typically ranging from 0.1 mL to 10 mL or larger used in a spectrophotometer. For the purposes of spectrophotometry (i.e. when making calculations using the Beer-Lambert law) the _________ is measured in centimeters (rather than in meters).

   a. Phosphorescence
   b. Photoacoustic spectroscopy
   c. Photothermal microspectroscopy
   d. Path length

----

5. In particle physics, wave mechanics and optics, _________ is the amount of momentum that one particle gives to another particle.

In the simplest example of scattering of two colliding particles with initial momenta $\vec{p}_{i1}, \vec{p}_{i2}$, resulting in final momenta $\vec{p}_{f1}, \vec{p}_{f2}$, the _________ is given by

$$\vec{q} = \vec{p}_{i1} - \vec{p}_{f1} = \vec{p}_{f2} - \vec{p}_{i2}$$

where the last identity expresses momentum conservation. _________ is an important quantity because $\Delta x = \hbar/|q|$ is a better measure for the typical distance resolution of the reaction than the momenta themselves.

a. Neutron depth profiling
b. Neutron howitzer
c. Momentum transfer
d. Neutron scanner

ANSWER KEY

Chapter 6. Interactions of Charged Particles with Matter

1. c
2. d
3. a
4. d
5. c

CHAPTER OUTLINE: KEY TERMS, PEOPLE, PLACES, CONCEPTS
Chapter 7
Interactions of Photons with Matter

______ Pair production

______ Photodisintegration

______ Photoelectric effect

______ Rayleigh

______ Rayleigh scattering

______ Thomson scattering

______ Attenuation

______ Attenuation coefficient

______ Exponential function

______ Mean free path

______ Mass attenuation coefficient

______ Photon

______ Compton wavelength

______ Wavelength

______ Law of cosines

______ Compton scattering

______ Charged particle

______ Binding energy

______ Equivalent

Chapter 7. Interactions of Photons with Matter

Momentum

Momentum transfer

Absorption edge

Fine structure

L-shell

Spectroscopic notation

Threshold energy

Born approximation

Classical electron radius

Annihilation

Positronium

Positron

Endothermic

Giant resonance

Induced radioactivity

Ionizing radiation

Bremsstrahlung

Pair production	Pair production refers to the creation of an elementary particle and its antiparticle, usually when a photon (or another neutral boson) interacts with a nucleus. For example an electron and its antiparticle, the positron, may be created. This is allowed, provided there is enough energy available to create the pair - at least the total rest mass energy of the two particles - and that the situation allows both energy and momentum to be conserved.
Photodisintegration	Photodisintegration is a physical process in which an extremely high energy gamma ray interacts with an atomic nucleus and causes it to enter an excited state, which immediately decays by emitting a subatomic particle. A single proton or neutron is effectively knocked out of the nucleus by the incoming gamma ray. This process is essentially the reverse of nuclear fusion, where lighter elements at high temperatures combine together forming heavier elements and releasing energy.
Photoelectric effect	In the photoelectric effect, electrons are emitted from matter (metals and non-metallic solids, liquids or gases) as a consequence of their absorption of energy from electromagnetic radiation of very short wavelength, such as visible or ultraviolet radiation. Electrons emitted in this manner may be referred to as photoelectrons. First observed by Heinrich Hertz in 1887, the phenomenon is also known as the Hertz effect, although the latter term has fallen out of general use.
Rayleigh	The rayleigh is a unit of photon flux, used to measure air glow (auroras, for example). It was first proposed in 1956 by D. M. Hunten, Franklin E. Roach, and J. W. Chamberlain. It is named for Robert John Sttt, 4th Baron Rayleigh (1875-1947).
Rayleigh scattering	Rayleigh scattering, is the elastic scattering of light or other electromagnetic radiation by particles much smaller than the wavelength of the light. The particles may be individual atoms or molecules. It can occur when light travels through transparent solids and liquids, but is most prominently seen in gases.
Thomson scattering	Thomson scattering is the elastic scattering of electromagnetic radiation by a free charged particle, as described by classical electromagnetism. It is just the low-energy limit of Compton scattering: the particle kinetic energy and photon frequency are the same before and after the scattering. This limit is valid as long as the photon energy is much less than the mass energy of the particle: $\nu \ll mc^2/h$.
Attenuation	In physics, attenuation is the gradual loss in intensity of any kind of flux through a medium. For instance, sunlight is attenuated by dark glasses, X-rays are attenuated by lead, and light and sound are attenuated by water.

In electrical engineering and telecommunications, attenuation affects the propagation of waves and signals in electrical circuits, in optical fibers, as well as in air (radio waves).

Attenuation coefficient	The attenuation coefficient is a quantity that charterizes how easily a material or medium can be penetrated by a beam of light, sound, particles, or other energy or matter. A large attenuation coefficient means that the beam is quickly 'attenuated' (weakened) as it passes through the medium, and a small attenuation coefficient means that the medium is relatively transparent to the beam. Attenuation coefficient is measured using units of reciprocal length.
Exponential function	In mathematics, the exponential function is the function e^x, where e is the number (approximately 2.718281828) such that the function e^x is its own derivative. The exponential function is used to model a relationship in which a constant change in the independent variable gives the same proportional change (i.e. percentage increase or decrease) in the dependent variable. The function is often written as exp(x), especially when it is impractical to write the independent variable as a superscript.
Mean free path	In physics, the mean free path is the average distance covered by a moving particle (such as an atom, a molecule, a photon) between successive impacts (collisions) which modify its direction or energy or other particle properties. Derivation Imagine a beam of particles being shot through a target, and consider an infinitesimally thin slab of the target. The atoms (or particles) that might stop a beam particle are shown in red.
Mass attenuation coefficient	The mass attenuation coefficient is a measurement of how strongly a chemical species or substance absorbs or scatters light at a given wavelength, per unit mass. In addition to visible light, mass attenuation coefficients can be defined for other electromagnetic radiation (such as X-rays), sound, or any other beam that attenuates. The mass attenuation coefficient is also called mass absorption coefficient or mass extinction coefficient.

Photon

In physics, a photon is an elementary particle, the quantum of light and all other forms of electromagnetic radiation, and the force carrier for the electromagnetic force. The effects of this force are easily observable at both the microscopic and macroscopic level, because the photon has no rest mass; this allows for interactions at long distances. Like all elementary particles, photons are currently best explained by quantum mechanics and exhibit wave-particle duality, exhibiting properties of both waves and particles.

Compton wavelength

The Compton wavelength is a quantum mechanical property of a particle. It was introduced by Arthur Compton in his explanation of the scattering of photons by electrons (a process known as Compton scattering). The Compton wavelength of a particle is equivalent to the wavelength of a photon whose energy is the same as the rest-mass energy of the particle.

Wavelength

In physics, the wavelength of a sinusoidal wave is the spatial period of the wave--the distance over which the wave's shape repeats. It is usually determined by considering the distance between consecutive corresponding points of the same phase, such as crests, troughs, or zero crossings, and is a characteristic of both traveling waves and standing waves, as well as other spatial wave patterns. Wavelength is commonly designated by the Greek letter lambda (λ).

Law of cosines

In trigonometry, the law of cosines relates the lengths of the sides of a plane triangle to the cosine of one of its angles. Using notation as in Fig. 1, the law of cosines says

$$c^2 = a^2 + b^2 - 2ab\cos\gamma$$

where γ denotes the angle contained between sides of lengths a and b and opposite the side of length c.

Some schools also describe the notation as follows:

$$c^2 = a^2 + b^2 - 2ab\cos C$$

Where C represents the same as γ and the rest of the parameters are the same.

Chapter 7. Interactions of Photons with Matter

Compton scattering	In physi, Compton scattering is a type of scattering that X-rays and gamma rays (both photons with different energy ranges) undergo in matter. The inelastic scattering of photons in matter results in a decrease in energy (increase in wavelength) of an X-ray or gamma ray photon, called the Compton effect. Part of the energy of the X/gamma ray is transferred to a scattering electron, which recoils and is ejected from its atom (which becomes ionized), and the rest of the energy is taken by the scattered, 'degraded' photon.
Charged particle	In physics, a charged particle is a particle with an electric charge. It may be either a subatomic particle or an ion. A collection of charged particles, or even a gas containing a proportion of charged particles, is called a plasma, which is called the fourth state of matter because its properties are quite different from solids, liquids and gases (plasma is the most common state of matter in the universe).
Binding energy	Binding energy is the mechanical energy required to disassemble a whole into separate parts. A bound system typically has a lower potential energy than its constituent parts; this is what keeps the system together--often this means that energy is released upon the creation of a bound state. The usual convention is that this corresponds to a positive binding energy.
Equivalent	The equivalent (symbol: eq or Eq), sometimes termed the molar equivalent, is a unit of amount of substance used in chemistry and the biological sciences. The equivalent is formally defined as the amount of a substance which will either: • react with or supply one mole of hydrogen ions (H^+) in an acid-base reaction; or • react with or supply one mole of eltrons in a redox reaction. The mass of one equivalent of a substance is called its equivalent weight. A historical definition, used espially for the chemical elements, describes an equivalent as the amount of a substance that will react with one gram of hydrogen, or with eight grams of oxygen, or with 35.5 grams (1.25 oz) of chlorine, or displaces any of the three.

Momentum

In classical mechanics, linear momentum or translational momentum is the product of the mass and velocity of an object:

$$\mathbf{p} \equiv m\mathbf{v}.$$

Like velocity, linear momentum is a vector quantity, possessing a direction as well as a magnitude. Linear momentum is also a conserved quantity, meaning that if a closed system is not affected by external forces, its total linear momentum cannot change.

Momentum transfer

In particle physics, wave mechanics and optics, momentum transfer is the amount of momentum that one particle gives to another particle.

In the simplest example of scattering of two colliding particles with initial momenta $\vec{p}_{i1}, \vec{p}_{i2}$, resulting in final momenta $\vec{p}_{f1}, \vec{p}_{f2}$, the momentum transfer is given by

$$\vec{q} = \vec{p}_{i1} - \vec{p}_{f1} = \vec{p}_{f2} - \vec{p}_{i2}$$

where the last identity expresses momentum conservation. Momentum transfer is an important quantity because $\Delta x = \hbar/|q|$ is a better measure for the typical distance resolution of the reaction than the momenta themselves.

Absorption edge

An absorption edge, adsorption discontinuity or adsorption limit is a sharp discontinuity in the absorption spectrum of a substance. These discontinuities occur at wavelengths where the energy of an absorbed photon corresponds to an electronic transition or ionization potential. When the quantum energy of the incident radiation becomes smaller than the work required to eject an electron from one or other quantum states in the constituent absorbing atom the incident radiation ceases to be absorbed by that state.

Fine structure

In atomic physics, the fine structure describes the splitting of the spectral lines of atoms due to first order relativistic corrections.

	The gross structure of line spectra is the line spectra predicted by non-relativistic electrons with no spin. For a hydrogenic atom, the gross structure energy levels only depend on the principal quantum number n.
L-shell	The L-shell, L-value, or McIlwain L-parameter (after Carl McIlwain) is a parameter describing a particular set of planetary magnetic field lines. Colloquially, L-value often describes the set of magnetic field lines which cross the Earth's magnetic equator at a number of Earth-radii equal to the L-value. For example, '$L = 2$' describes the set of the Earth's magnetic field lines which cross the Earth's magnetic equator two earth radii from the center of the Earth.
Spectroscopic notation	Spectroscopic notation provides various ways to specify atomic ionization states, as well as atomic and molecular orbitals. Ionization states Spectroscopists customarily refer to the spectrum arising from a given ionization state of a given element by the element's symbol followed by a Roman numeral. The numeral I is used for spectral lines associated with the neutral element, II for those from the first ionization state, III for those from the second ionization state, and so on.
Threshold energy	In particle physics, the threshold energy for production of a particle is the minimum kinetic energy a pair of traveling particles must have when they collide. The threshold energy is always grear than or equal to the rest energy of the desired particle. In most cases, since momentum is also conserved, the threshold energy is significantly grear than the rest energy of the desired particle - and thus there will still be considerable kinetic energy in the final particles.
Born approximation	In scattering theory and, in particular in quantum mechanics, the Born approximation consists of taking the incident field in place of the total field as the driving field at each point in the scatterer. Born approximation is named after Max Born, winner of the 1954 Nobel Prize for physics. It is the perturtion method applied to scattering by an extended body.

Classical electron radius	The classical electron radius, is based on a classical (i.e., non-quantum) relativistic model of the electron. Its value is calculated as

$$r_e = \frac{1}{4\pi\varepsilon_0}\frac{e^2}{m_e c^2} = 2.8179402894(58) \times 10^{-15}\text{m}$$

where e and m_e are the electric charge and the mass of the electron, c is the speed of light, and ε_0 is the permittivity of free space.

In cgs units, this becomes more simply

$$r_e = \frac{e^2}{m_e c^2} = 2.8179402894(58) \times 10^{-13}\text{cm}$$

with (to three significant digits)

$$e = 4.80 \times 10^{-10}\text{esu}, m_e = 9.11 \times 10^{-28}\text{g}, c = 3.00 \times 10^{10}\text{cm/s}.$$

Using classical electrostatics, the energy required to assemble a sphere of constant charge density, of radius r_e and charge e is

$$E = \frac{3}{5}\frac{1}{4\pi\varepsilon_0}\frac{e^2}{r_e}.$$

If the charge is on the surface the energy is

$$E = \frac{1}{2}\frac{1}{4\pi\varepsilon_0}\frac{e^2}{r_e}.$$

Ignoring the factors 3/5 or 1/2, if this is equated to the relativistic energy of the electron ($E = mc^2$) and solved for r_e , the above result is obtained.

Annihilation

Annihilation is defined as 'total destruction' or 'complete obliteration' of an object; having its root in the Latin nihil (nothing). A literal translation is 'to make into nothing'.

In physics, the word is used to denote the process that occurs when a subatomic particle collides with its respective antiparticle.

Positronium

Positronium is a system consisting of an electron and its anti-particle, a positron, bound together into an 'exotic atom'. Being unstable, the two particles annihilate each other to produce two gamma ray photons after an average lifetime of 125 ps or three gamma ray photons after 142 ns in vacuum, depending on the relative spin states of the positron and electron. The orbit of the two particles and the set of energy levels is similar to that of the hydrogen atom (electron and proton).

Positron

The positron is the antiparticle or the antimatter counterpart of the electron. The positron has an electric charge of +1e, a spin of 1/2, and has the same mass as an electron. When a low-energy positron collides with a low-energy electron, annihilation occurs, resulting in the production of two or more gamma ray photons .

Endothermic

In thermodynamics, the word endothermic describes a process or reaction in which the system absorbs energy from the surroundings in the form of heat. It is a modern coinage formed from Greek roots (as is often the case with scientific terminology). The prefix endo- derives from the Greek word 'endon' (?νδον) meaning 'within,' and the latter part of the word comes from the Greek word root 'therm' (θερμ-) meaning 'hot.' Hence it refers to a reaction that needs heat.

Giant resonance

Giant resonance is a high-frequency collective excitation of atomic nuclei, as a property of many-body quantum systems. In the macroscopic interpretation of such an excitation in terms of an oscillation, the most prominent giant resonance is a collective oscillation of all protons against all neutrons in a nucleus.

1947, G. C. Baildwin and G. S. Klaiber observed the giant dipole resonance (GDR) in photonuclear reactions, in 1972 the giant quadrupole resonance (GQR) was discovered, and in 1977 the giant monopole resonance (GMR) was discovered in medium and heavy nuclei.

Induced radioactivity | Induced radioactivity occurs when a previously stable material has been made radioactive by exposure to specific radiation. Most radioactivity does not induce other material to become radioactive.

Neutron activation is the main form of induced radioactivity, which happens when free neutrons are captured by nuclei.

Ionizing radiation | Ionizing (or ionising) radiation is radiation composed of particles that individually can liberate an electron from an atom or molecule, producing ions, which are atoms or molecules with a net electric charge. These tend to be especially chemically reactive, and the reactivity produces the high biological damage caused per unit of energy of ionizing radiation.

The degree and nature of such ionization depends on the energy and type of the individual particles composing the radiation, and less upon the radiation particle number.

Bremsstrahlung | Bremsstrahlung is electromagnetic radiation produced by the deceleration of a charged particle when deflected by another charged particle, typically an electron by an atomic nucleus. The moving particle loses kinetic energy, which is converted into a photon because energy is conserved. The term is also used to refer to the process of producing the radiation.

PRACTICE QUIZ

Chapter 7. Interactions of Photons with Matter

1. In physi, _________ is a type of scattering that X-rays and gamma rays (both photons with different energy ranges) undergo in matter. The inelastic scattering of photons in matter results in a decrease in energy (increase in wavelength) of an X-ray or gamma ray photon, called the Compton effect. Part of the energy of the X/gamma ray is transferred to a scattering electron, which recoils and is ejected from its atom (which becomes ionized), and the rest of the energy is taken by the scattered, 'degraded' photon.

   a. Compton wavelength
   b. Double-slit experiment
   c. Hanle effect
   d. Compton scattering

2. In atomic physics, the _________ describes the splitting of the spectral lines of atoms due to first order relativistic corrections.

   The gross structure of line spectra is the line spectra predicted by non-relativistic electrons with no spin. For a hydrogenic atom, the gross structure energy levels only depend on the principal quantum number n.

   a. Free-fall atomic model
   b. G-factor
   c. Geonium atom
   d. Fine structure

3. The _________(symbol: eq or Eq), sometimes termed the molar equivalent, is a unit of amount of substance used in chemistry and the biological sciences.

   The equivalent is formally defined as the amount of a substance which will either:

   - react with or supply one mole of hydrogen ions (H^+) in an acid-base reaction; or
   - react with or supply one mole of eltrons in a redox reaction.

   The mass of one equivalent of a substance is called its equivalent weight.

A historical definition, used espially for the chemical elements, describes an equivalent as the amount of a substance that will react with one gram of hydrogen, or with eight grams of oxygen, or with 35.5 grams (1.25 oz) of chlorine, or displaces any of the three.

a. Equivalent weight
b. Equivalent
c. Charged-particle equilibrium
d. Chemical shift

4. In classical mechanics, linear _________ or translational _________ is the product of the mass and velocity of an object:

$$\mathbf{p} \equiv m\mathbf{v}.$$

Like velocity, linear _________ is a vector quantity, possessing a direction as well as a magnitude. Linear _________ is also a conserved quantity, meaning that if a closed system is not affected by external forces, its total linear _________ cannot change.

a. Monogenic system
b. Motion
c. Momentum
d. Pieter van Musschenbroek

5. In the _________, electrons are emitted from matter (metals and non-metallic solids, liquids or gases) as a consequence of their absorption of energy from electromagnetic radiation of very short wavelength, such as visible or ultraviolet radiation. Electrons emitted in this manner may be referred to as photoelectrons. First observed by Heinrich Hertz in 1887, the phenomenon is also known as the Hertz effect, although the latter term has fallen out of general use.

a. Photoelectric effect
b. X-ray burster
c. Parity
d. Particle-induced X-ray emission

ANSWER KEY
Chapter 7. Interactions of Photons with Matter

---

1. d
2. d
3. b
4. c
5. a

CHAPTER OUTLINE: KEY TERMS, PEOPLE, PLACES, CONCEPTS

Chapter 8

Energy Transfer and Energy Absorption in Photon Interactions

___________ Mass attenuation coefficient

___________ Attenuation

___________ Attenuation coefficient

___________ Compton scattering

___________ Pair production

___________ Photoelectric effect

___________ Rayleigh

___________ Rayleigh scattering

___________ Absorption edge

___________ Weighted mean

___________ Charged particle

___________ Photon

___________ Annihilation

___________ Bremsstrahlung

___________ Ionization

___________ Atomic weight

___________ Water mass

___________ Photodisintegration

___________ Threshold energy

Atomic number

Delta ray

Nuclear reactor

CHAPTER HIGHLIGHTS: KEY TERMS, PEOPLE, PLACES, CONCEPTS
Chapter 8. Energy Transfer and Energy Absorption in Photon Interactions

Mass attenuation coefficient	The mass attenuation coefficient is a measurement of how strongly a chemical species or substance absorbs or scatters light at a given wavelength, per unit mass. In addition to visible light, mass attenuation coefficients can be defined for other electromagnetic radiation (such as X-rays), sound, or any other beam that attenuates. The mass attenuation coefficient is also called mass absorption coefficient or mass extinction coefficient.
Attenuation	In physics, attenuation is the gradual loss in intensity of any kind of flux through a medium. For instance, sunlight is attenuated by dark glasses, X-rays are attenuated by lead, and light and sound are attenuated by water. In electrical engineering and telecommunications, attenuation affects the propagation of waves and signals in electrical circuits, in optical fibers, as well as in air (radio waves).
Attenuation coefficient	The attenuation coefficient is a quantity that charterizes how easily a material or medium can be penetrated by a beam of light, sound, particles, or other energy or matter. A large attenuation coefficient means that the beam is quickly 'attenuated' (weakened) as it passes through the medium, and a small attenuation coefficient means that the medium is relatively transparent to the beam. Attenuation coefficient is measured using units of reciprocal length.
Compton scattering	In physi, Compton scattering is a type of scattering that X-rays and gamma rays (both photons with different energy ranges) undergo in matter. The inelastic scattering of photons in matter results in a decrease in energy (increase in wavelength) of an X-ray or gamma ray photon, called the Compton effect. Part of the energy of the X/gamma ray is transferred to a scattering electron, which recoils and is ejected from its atom (which becomes ionized), and the rest of the energy is taken by the scattered, 'degraded' photon.
Pair production	Pair production refers to the creation of an elementary particle and its antiparticle, usually when a photon (or another neutral boson) interacts with a nucleus. For example an electron and its antiparticle, the positron, may be created. This is allowed, provided there is enough energy available to create the pair - at least the total rest mass energy of the two particles - and that the situation allows both energy and momentum to be conserved.

Photoelectric effect	In the photoelectric effect, electrons are emitted from matter (metals and non-metallic solids, liquids or gases) as a consequence of their absorption of energy from electromagnetic radiation of very short wavelength, such as visible or ultraviolet radiation. Electrons emitted in this manner may be referred to as photoelectrons. First observed by Heinrich Hertz in 1887, the phenomenon is also known as the Hertz effect, although the latter term has fallen out of general use.
Rayleigh	The rayleigh is a unit of photon flux, used to measure air glow (auroras, for example). It was first proposed in 1956 by D. M. Hunten, Franklin E. Roach, and J. W. Chamberlain. It is named for Robert John Sttt, 4th Baron Rayleigh (1875-1947).
Rayleigh scattering	Rayleigh scattering, is the elastic scattering of light or other electromagnetic radiation by particles much smaller than the wavelength of the light. The particles may be individual atoms or molecules. It can occur when light travels through transparent solids and liquids, but is most prominently seen in gases.
Absorption edge	An absorption edge, adsorption discontinuity or adsorption limit is a sharp discontinuity in the absorption spectrum of a substance. These discontinuities occur at wavelengths where the energy of an absorbed photon corresponds to an electronic transition or ionization potential. When the quantum energy of the incident radiation becomes smaller than the work required to eject an electron from one or other quantum states in the constituent absorbing atom the incident radiation ceases to be absorbed by that state.
Weighted mean	The weighted mean is similar to an arithmetic mean (the most common type of average), where instead of each of the data points contributing equally to the final average, some data points contribute more than others. The notion of weighted mean plays a role in descriptive statistics and also occurs in a more general form in several other areas of mathematics. If all the weights are equal, then the weighted mean is the same as the arithmetic mean.
Charged particle	In physics, a charged particle is a particle with an electric charge. It may be either a subatomic particle or an ion. A collection of charged particles, or even a gas containing a proportion of charged particles, is called a plasma, which is called the fourth state of matter because its properties are quite different from solids, liquids and gases (plasma is the most common state of matter in the universe).

Chapter 8. Energy Transfer and Energy Absorption in Photon Interactions

Photon	In physics, a photon is an elementary particle, the quantum of light and all other forms of electromagnetic radiation, and the force carrier for the electromagnetic force. The effects of this force are easily observable at both the microscopic and macroscopic level, because the photon has no rest mass; this allows for interactions at long distances. Like all elementary particles, photons are currently best explained by quantum mechanics and exhibit wave-particle duality, exhibiting properties of both waves and particles.
Annihilation	Annihilation is defined as 'total destruction' or 'complete obliteration' of an object; having its root in the Latin nihil (nothing). A literal translation is 'to make into nothing'. In physics, the word is used to denote the process that occurs when a subatomic particle collides with its respective antiparticle.
Bremsstrahlung	Bremsstrahlung is electromagnetic radiation produced by the deceleration of a charged particle when deflected by another charged particle, typically an electron by an atomic nucleus. The moving particle loses kinetic energy, which is converted into a photon because energy is conserved. The term is also used to refer to the process of producing the radiation.
Ionization	Ionization is the process of converting an atom or molecule into an ion by adding or removing charged particles such as electrons or other ions. A substance may dissociate without necessarily producing ions. As an example, the molecules of table sugar dissociate in water (sugar is dissolved) but exist as intact neutral entities.
Atomic weight	Atomic weight is a dimensionless physical quantity, the ratio of the average mass of atoms of an element (from a given source) to 1/12 of the mass of an atom of carbon-12 (known as the unified atomic mass unit). The term is usually used, without further qualification, to refer to the standard atomic weights published at regular intervals by the International Union of Pure and Applied Chemistry (IUPAC) and which are intended to be applicable to normal laboratory materials.
Water mass	An oceanographic water mass is an identifiable body of water with a common formation history which has physical properties distinct from surrounding water. Properties include temperature, salinity, chemical - isotopic ratios, and other physical quantities.

	Common water masses in the world ocean are: Antarctic Bottom Water (AABW), North Atlantic Deep Water (NADW), Circumpolar Deep Water (CDW), Antarctic Intermediate Water (AAIW), Subantarctic Mode Water (SAMW), Arctic Intermediate Water (AIW), the central waters of various oceanic basins, and various surface waters.
Photodisintegration	Photodisintegration is a physical process in which an extremely high energy gamma ray interacts with an atomic nucleus and causes it to enter an excited state, which immediately decays by emitting a subatomic particle. A single proton or neutron is effectively knocked out of the nucleus by the incoming gamma ray. This process is essentially the reverse of nuclear fusion, where lighter elements at high temperatures combine together forming heavier elements and releasing energy.
Threshold energy	In particle physics, the threshold energy for production of a particle is the minimum kinetic energy a pair of traveling particles must have when they collide. The threshold energy is always grear than or equal to the rest energy of the desired particle. In most cases, since momentum is also conserved, the threshold energy is significantly grear than the rest energy of the desired particle - and thus there will still be considerable kinetic energy in the final particles.
Atomic number	In chemistry d physics, the atomic number is the number of protons found in the nucleus of atom d therefore identical to the charge number of the nucleus. It is conventionally represented by the symbol Z. The atomic number uniquely identifies a chemical element. In atom of neutral charge, the atomic number is also equal to the number of electrons.
Delta ray	A delta ray is sometimes used to describe any recoil particle caused by secondary ionization. The term was coined by J.J. Thomson. It is entirely unrelated to the family of subatomic particles named delta baryon.
Nuclear reactor	A nuclear reactor is a device to initiate and control a sustained nuclear chain reaction. Most commonly they are used for generating electricity and for the propulsion of ships. Usually heat from nuclear fission is passed to a working fluid (water or gas), which runs through turbines that power either ship's propellers or generators.

PRACTICE QUIZ

Chapter 8. Energy Transfer and Energy Absorption in Photon Interactions

---

1. The _________ is similar to an arithmetic mean (the most common type of average), where instead of each of the data points contributing equally to the final average, some data points contribute more than others. The notion of _________ plays a role in descriptive statistics and also occurs in a more general form in several other areas of mathematics.

   If all the weights are equal, then the _________ is the same as the arithmetic mean.

   a. Whitney extension theorem
   b. Actinism
   c. Activation product
   d. Weighted mean

2. _________ is electromagnetic radiation produced by the deceleration of a charged particle when deflected by another charged particle, typically an electron by an atomic nucleus. The moving particle loses kinetic energy, which is converted into a photon because energy is conserved. The term is also used to refer to the process of producing the radiation.

   a. Brillouin scattering
   b. Chaotic scattering
   c. Core-excited shape resonance
   d. Bremsstrahlung

3. The _________ is a measurement of how strongly a chemical species or substance absorbs or scatters light at a given wavelength, per unit mass. In addition to visible light, _________s can be defined for other electromagnetic radiation (such as X-rays), sound, or any other beam that attenuates.

   The _________ is also called mass absorption coefficient or mass extinction coefficient.

   a. Mass flux
   b. Maximum density
   c. Mechanical impedance
   d. Mass attenuation coefficient

----

4. _________ is a dimensionless physical quantity, the ratio of the average mass of atoms of an element (from a given source) to 1/12 of the mass of an atom of carbon-12 (known as the unified atomic mass unit). The term is usually used, without further qualification, to refer to the standard _________s published at regular intervals by the International Union of Pure and Applied Chemistry (IUPAC) and which are intended to be applicable to normal laboratory materials.

   a. Equivalent
   b. Atomic weight
   c. Core-excited shape resonance
   d. Coulomb collision

5. An _________, adsorption discontinuity or adsorption limit is a sharp discontinuity in the absorption spectrum of a substance. These discontinuities occur at wavelengths where the energy of an absorbed photon corresponds to an electronic transition or ionization potential. When the quantum energy of the incident radiation becomes smaller than the work required to eject an electron from one or other quantum states in the constituent absorbing atom the incident radiation ceases to be absorbed by that state.

   a. Absorption spectroscopy
   b. Actinism
   c. Activation product
   d. Absorption edge

ANSWER KEY

Chapter 8. Energy Transfer and Energy Absorption in Photon Interactions

1. d

2. d

3. d

4. b

5. d

CHAPTER OUTLINE: KEY TERMS, PEOPLE, PLACES, CONCEPTS
Chapter 9
Interactions of Neutrons with Matter

______ Neutron

______ Neutron capture

______ Nuclear reaction

______ Boron neutron capture therapy

______ External

______ Neutron activation

______ Nuclear reactor

______ Elastic scattering

______ Inelastic scattering

______ Spallation

______ Nuclear fission

______ Cyclotron

______ Neutron generator

______ Generator

______ Binding energy

______ Neutron source

______ Neutron activation analysis

______ Specific activity

______ Neutron radiography

Multileaf collimator

Collimator

Photon

Neutron	The neutron is a subatomic hadron particle which has the symbol n or n^0, no net electric charge and a mass slightly larger than that of a proton. With the exception of hydrogen, nuclei of atoms consist of protons and neutrons, which are therefore collectively referred to as nucleons. The number of protons in a nucleus is the atomic number and defines the type of element the atom forms.
Neutron capture	Neutron capture is a kind of nuclear reaction in which an atomic nucleus collides with one or more neutrons and they merge to form a heavier nucleus. Sie neutrons have no electric charge they can enter a nucleus more easily than positively charged protons, which are repelled electrostatically. Neutron capture plays an important role in the cosmic nucleosynthesis of heavy elements.
Nuclear reaction	In nuclear physics and nuclear chemistry, a nuclear reaction is semantically considered to be the process in which two nuclei, or else a nucleus of an atom and a subatomic particle (such as a proton, neutron, or high energy electron) from outside the atom, collide to produce products different from the initial particles. In principle, a reaction can involve more than three particles colliding, but because the probability of three or more nuclei to meet at the same time at the same place is much less than for two nuclei, such an event is exceptionally rare. Radioactive decays can be considered to be spontaneous nuclear reactions, in as much as there is such a thing as a spontaneous chemical reaction.
Boron neutron capture therapy	Boron neutron capture therapy is an experimental form of radiotherapy that uses a neutron beam that interacts with boron injected into a patient. depends on the interaction of slow neutrons with boron-10 to produce alpha particles and lithium nuclei, without producing other types of ionizing radiation. Patients are first given an intravenous injection of a boron-10 containing chemical that preferentially binds to tumor cells.

External

The term external is useful for describing certain algebraic structures. The term comes from the concept of an external binary operation which is a binary operation that draws from some external set. To be more specific, a left external binary operation on S over R is a function $f : R \times S \to S$ and a right external binary operation on S over R is a function $f : S \times R \to S$ where S is the set the operation is defined on, and R is the external set (the set the operation is defined over).

Neutron activation

Neutron activation is the process in which neutron radiation induces radioactivity in materials, and occurs when atomic nuclei capture free neutrons, becoming heavier and entering excited states. The excited nucleus often decays immediately by emitting particles such as neutrons, protons, or alpha particles. The neutron capture, even after any intermediate decay, often results in the formation of an unstable activation product.

Nuclear reactor

A nuclear reactor is a device to initiate and control a sustained nuclear chain reaction. Most commonly they are used for generating electricity and for the propulsion of ships. Usually heat from nuclear fission is passed to a working fluid (water or gas), which runs through turbines that power either ship's propellers or generators.

Elastic scattering

In scattering theory and in particular in particle physics, elastic scattering is one of the specific forms of scattering. In this procs, the kinetic energy of the incident particl is conserved, only their direction of propagation is modified (by interaction with other particl and/or a potential).

Electron elastic scattering
When an alpha particle is an incident particle and it is diffracted in the Coulomb potential of atoms and molecul, the elastic scattering procs is called Rutherford scattering.

Inelastic scattering

In particle physics and chemtry, inelastic scattering a fundamental scattering process in which the kinetic energy of an incident particle not conserved (in contrast to elastic scattering). In an inelastic scattering process, some of the energy of the incident particle lost or gained. Although the term htorically related to the concept of inelastic collion in dynamics, the two concepts are quite dtinct; the latter refers to processes in which the total kinetic energy not conserved.

Spallation

In general, spallation is a process in which fragments of material (spall) are ejected from a body due to impact or stress. In the context of impact mechanics it describes ejection or vaporization of material from a target during impact by a projectile. In planetary physics, spallation describes meteoritic impacts on a planetary surface and the effects of a stellar wind on a planetary atmosphere.

Chapter 9. Interactions of Neutrons with Matter

Nuclear fission	In nuclear physics and nuclear chemistry, nuclear fission refers to either a nuclear reaction or a radioactive decay process in which the nucleus of an atom splits into smaller parts (lighter nuclei), often producing free neutrons and photons (in the form of gamma rays), and releasing a very large amount of energy, even by the energetic standards of radioactive decay. The two nuclei produced are most often of comparable but slightly different sizes, typically with a mass ratio of products of about 3 to 2, for common fissile isotopes. Most fissions are binary fissions (producing two charged fragments), but occasionally (2 to 4 times per 1000 events), three positively charged fragments are produced, in a ternary fission.
Cyclotron	A cyclotron is a compact type of particle accelerator in which charged particles in a static magnetic field are travelling outwards from the center along a spiral path and get accelerated by radio frequency electromagnetic fields. The cyclotron was invented by Leó Szilárd and first manufactured and patented by Ernest Lawrence, of the University of California, Berkeley who started operating it in 1932 The first European cyclotron was constructed in Leningrad in the physics department of the Radium Institute, headed by Vitali Khlopin.
Neutron generator	Neutron generators are neutron source devices which contain compact linear accelerators and that produce neutrons by fusi isotopes of hydrogen together. The fusion reactions take place in these devices by accelerati either deuterium, tritium, or a mixture of these two isotopes into a metal hydride target which also contains either deuterium, tritium or a mixture. Fusion of deuterium atoms (D + D) results in the formation of a He-3 ion and a neutron with a kinetic energy of approximately 2.5 MeV. Fusion of a deuterium and a tritium atom (D + T) results in the formation of a He-4 ion and a neutron with a kinetic energy of approximately 14.1 MeV. Thousands of such small, relatively inexpensive systems have been built over the past five decades.
Generator	In category theory in mathematics a generator (or separator) of a category C is an object G of the category, such that for any two different morphisms $f, g : X \to Y$ in C, there is a morphism $h : G \to X$, such that the compositions $f \circ h \neq g \circ h$.

Generators are central to the definition of Grothendieck categories.

Examples

- In the category of abelian groups, the group of integers $\mathbf{Z}$ is a generator: If f and g are different, then there is an element $x \in X$, such that $f(x) \neq g(x)$.

Binding energy

Binding energy is the mechanical energy required to disassemble a whole into separate parts. A bound system typically has a lower potential energy than its constituent parts; this is what keeps the system together--often this means that energy is released upon the creation of a bound state. The usual convention is that this corresponds to a positive binding energy.

Neutron source

A Neutron source is a device that emits neutro. There are a wide variety of different sources, ranging from hand-held radioactive sources to neutron research facilities operating research reactors and spallation sources. Depending upon neutron energy, neutron flux, size of the source, costs, and government regulatio, these devices find use in a diverse array of applicatio in areas of physics, engineering, medicine, nuclear weapo, petroleum exploration, biology, chemistry, nuclear power and other industries.

Neutron activation analysis

In chemistry, neutron activation analysis is a nuclear process used for determining the concentrations of elements in a vast amount of materials. allows discrete sampling of elements as it disregards the chemical form of a sample, and focuses solely on its nucleus. The method is based on neutron activation and therefore requires a source of neutrons.

Specific activity

In nuclear sciences and technologies, 'activity' is the SI quantity related to the phenomenon of natural and artificial radioactivity. The SI unit of 'activity' is becquerel, Bq, while that of 'specific activity' is Bq/kg. The old unit of 'activity' was curie, Ci, while that of 'specific activity' was Ci/g.

Neutron radiography

Neutron radiography is the process by which film is exposed by first passing neutrons through an object to produce a visible image of the materials that make up the object. It is primarily used in scientific investigations.

History

The neutron was discovered by James Chadwick in 1932. The first demonstration of neutron radiography was made by Hartmut Kallmann and E. Kuhn in the late 1930s; they discovered that upon bombardment with neutrons, some materials emitted radiation that could expose a photographic film.

Multileaf collimator

A multileaf collimator is a device made up of individual 'leaves' of a high atomic numbered material, usually tungsten, that can move independently in and out of the path of a particle beam in order to block it.

MLCs are used on linear accelerators to provide conformal shaping of radiotherapy treatment beams. Specifically, conformal radiotherapy and Intensity Modulated Radiation Therapy (IMRT) can be delivered using MLC's.

Collimator

A collimator is a device that narrows a beam of particles or waves. To 'narrow' can mean either to cause the directions of motion to become more aligned in a specific direction (i.e., collimated or parallel) or to cause the spatial cross section of the beam to become smaller.

Optical collimators
In optics, a collimator may consist of a curved mirror or lens with some type of light source and/or an image at its focus.

Photon

In physics, a photon is an elementary particle, the quantum of light and all other forms of electromagnetic radiation, and the force carrier for the electromagnetic force. The effects of this force are easily observable at both the microscopic and macroscopic level, because the photon has no rest mass; this allows for interactions at long distances. Like all elementary particles, photons are currently best explained by quantum mechanics and exhibit wave-particle duality, exhibiting properties of both waves and particles.

----

1. A _________ is a device made up of individual 'leaves' of a high atomic numbered material, usually tungsten, that can move independently in and out of the path of a particle beam in order to block it.

MLCs are used on linear accelerators to provide conformal shaping of radiotherapy treatment beams. Specifically, conformal radiotherapy and Intensity Modulated Radiation Therapy (IMRT) can be delivered using MLC's.

a. Non-ionizing radiation
b. Particle therapy
c. Percentage depth dose curve
d. Multileaf collimator

2. A _________ is a compact type of particle accelerator in which charged particles in a static magnetic field are travelling outwards from the center along a spiral path and get accelerated by radio frequency electromagnetic fields.

The _________ was invented by Leó Szilárd and first manufactured and patented by Ernest Lawrence, of the University of California, Berkeley who started operating it in 1932 The first European _________ was constructed in Leningrad in the physics department of the Radium Institute, headed by Vitali Khlopin.

a. Cyclotron resonance
b. Cyclotron
c. Dipole magnet
d. Dynamitron

3. In general, _________ is a process in which fragments of material (spall) are ejected from a body due to impact or stress. In the context of impact mechanics it describes ejection or vaporization of material from a target during impact by a projectile. In planetary physics, _________ describes meteoritic impacts on a planetary surface and the effects of a stellar wind on a planetary atmosphere.

a. Spent nuclear fuel shipping cask
b. Steel plate construction
c. Stirling radioisotope generator
d. Spallation

----

4. In chemistry, _________ is a nuclear process used for determining the concentrations of elements in a vast amount of materials. allows discrete sampling of elements as it disregards the chemical form of a sample, and focuses solely on its nucleus. The method is based on neutron activation and therefore requires a source of neutrons.

   a. Qualitative inorganic analysis
   b. Separation process
   c. Neutron activation analysis
   d. Nuclear electric rocket

5. In nuclear physics and nuclear chemistry, _________ refers to either a nuclear reaction or a radioactive decay process in which the nucleus of an atom splits into smaller parts (lighter nuclei), often producing free neutrons and photons (in the form of gamma rays), and releasing a very large amount of energy, even by the energetic standards of radioactive decay. The two nuclei produced are most often of comparable but slightly different sizes, typically with a mass ratio of products of about 3 to 2, for common fissile isotopes. Most fissions are binary fissions (producing two charged fragments), but occasionally (2 to 4 times per 1000 events), three positively charged fragments are produced, in a ternary fission.

   a. Nuclear transmutation
   b. Nucleogenic
   c. Nuclear fission
   d. Positron emission

ANSWER KEY
Chapter 9. Interactions of Neutrons with Matter

---

1. d
2. b
3. d
4. c
5. c

CHAPTER OUTLINE: KEY TERMS, PEOPLE, PLACES, CONCEPTS

Chapter 10

Kinetics of Radioactive Decay

- ________ Curie
- ________ Polonium
- ________ Radioactive decay
- ________ Radium
- ________ Radionuclide
- ________ Nuclide
- ________ Avogadro
- ________ Specific activity
- ________ Half-life
- ________ Decay chain
- ________ Mixture
- ________ Characteristic time
- ________ Transient
- ________ Transient equilibrium
- ________ Secular equilibrium
- ________ Branching fraction
- ________ Neutron

Curie	The curie is a non-SI unit of radioactivity. It is defined as $1\ \text{Ci} = 3.7 \times 10^{10}$ decays per second. Its continued use is discouraged. One curie is roughly the activity of 1 gram of the radium isotope ^{226}Ra, a substance studied by the Curies.
Polonium	Polonium is a chemical element with the symbol Po and atomic number 84, discovered in 1898 by Marie Sklodowska-Curie and Pierre Curie. A rare and highly radioactive element, polonium is chemically similar to bismuth and tellurium, and it occurs in uranium ores. Polonium has been studied for possible use in heating spacecraft.
Radioactive decay	Radioactive decay is the process by which an atomic nucleus of an unstable atom loses energy by emitting ionizing particles (ionizing radiation). The emission is spontaneous, in that the atom decays without any physical interaction with another particle from outside the atom. Usually, radioactive decay happens due to a process confined to the nucleus of the unstable atom, but, on occasion (as with the different processes of electron capture and internal conversion), an inner electron of the radioactive atom is also necessary to the process.
Radium	Radium is a chemical element with atomic number 88, represented by the symbol Ra. Radium is an almost pure-white alkaline earth metal, but it readily oxidizes on exposure to air, becoming black in color. All isotopes of radium are highly radioactive, with the most stable isotope being radium-226, which has a half-life of 1601 years and decays into radon gas.
Radionuclide	A radionuclide is an atom with an unstable nucleus, which is a nucleus characterized by excess energy available to be imparted either to a newly created radiation particle within the nucleus or to an atomic electron. The radionuclide, in this process, undergoes radioactive decay, and emits gamma ray(s) and/or subatomic particles. These particles constitute ionizing radiation.

Nuclide

A nuclide is an atomic species characterized by the specific constitution of its nucleus, i.e., by its number of protons Z, its number of neutrons N, and its nuclear energy state.

The word nuclide was proposed by Truman P. Kohman in 1947. Doctor Kohman originally suggested nuclide as referring to a 'species of nucleus' defined by containing a certain number of neutrons and protons. The word thus was originally intended to focus on the nucleus.

Avogadro

Avogadro (software) is a molecular editor designed for cross-platform use in computational chemistry, molecular modeling, bioinformatics, materials science, and related are. It is extensible through a plugin architecture.

Features

- Molecular builder/editor for Windows, Linux, and Mac OS X.
- All source code is available under the GNU GPL.
- Translations into Chinese, French, German, Italian, Russian, and Spanish.
- Supports multi-threaded rendering and computation.
- Plugin architecture for developers, including rendering, interactive tools, commands, and Python scripts.
- OpenBabel import of files, input generation for multiple computational chemistry packages, crystallography, and biomolecules.

.

Specific activity

In nuclear sciences and technologies, 'activity' is the SI quantity related to the phenomenon of natural and artificial radioactivity. The SI unit of 'activity' is becquerel, Bq, while that of 'specific activity' is Bq/kg. The old unit of 'activity' was curie, Ci, while that of 'specific activity' was Ci/g.

Half-life

Half-life, abbreviated $t_{1/2}$, is the period of time it takes for the amount of a substance undergoing decay to decrease by half. The name was originally used to describe a characteristic of unstable atoms (radioactive decay), but it may apply to any quantity which follows a set-rate decay.

The original term, dating to 1907, was 'half-life period', which was later shortened to 'half-life' in the early 1950s.

Decay chain

In nuclear science, the decay chain refers to the radioactive decay of different discrete radioactive decay products as a chained series of transformations. Most radioactive elements do not decay directly to a stable state, but rather undergo a series of decays until eventually a stable isotope is reached.

Decay stages are referred to by their relationship to previous or subsequent stages.

Mixture

In chemistry, a mixture is a material system made up by two or more different substances which are mixed but are not combined chemically. Mixture refers to the physical combination of two or more substances the identities of which are retained and are mixed in the form of alloys, solutions, suspensions, and colloids.

Mixtures are the product of a mechanical blending or mixing of chemical substances like elements and compounds, without chemical bonding or other chemical change, so that each ingredient substance retains its own chemical properties and makeup.

Characteristic time

The characteristic time is an estimate of the order of magnitude of the reaction time scale of a system. It can loosely be defined as the inverse of the reaction rate. In chemistry, the characteristic time is used to determine whether the problem needs to be solved as an equilibrium problem or a kinetic problem.

Transient

A transient event is a short-lived burst of energy in a system caused by a sudden change of state.

The source of the transient energy may be an internal event or a nearby event. The energy then couples other parts of the system, typically appearing as a short burst of oscillation.

Transient equilibrium	In nuclear physics, transient equilibrium is a situation in which equilibrium is reached by a parent-daughr radioactive isotope pair where the half-life of the daughr is shorr than the half-life of the parent. Contrary to Secular equilibrium, the half-life of the daughr is not negligible compare to parent's half-life. An example of this is the Molybdenum 99 generator producing chnetium 99 for nuclear medicine diagnostic procedures.
Secular equilibrium	In nuclear physics, Secular equilibrium is a situation in which the quantity of a radioactive isotope remains constant because its production rate (due, e.g., to decay of a parent isotope) is equal to its decay rate. Secular equilibrium can only occur in a radioactive decay chain if the half-life of the daughter radionuclide B is much shorter than the half-life of the parent radionuclide A. In such a situation, the decay rate of A, and hence the production rate of B, is approximately constant, because the half-life of A is very long compared to the timescales being considered. The quantity of radionuclide B builds up until the number of B atoms decaying per unit time becomes equal to the number being produced per unit time; the quantity of radionuclide B then reaches a constant, equilibrium value.
Branching fraction	In particle physics and nuclear physics, the branching fraction for a decay is the fraction of particles which decay by an individual decay mode with respect to the total number of particles which decay. It is equal to the ratio of the partial decay constant to the overall decay constant. Sometimes a partial half-life is given, but this term is misleading; due to competing modes it is not true that half of the particles will decay through a particular decay mode after its partial half-life.
Neutron	The neutron is a subatomic hadron particle which has the symbol n or n^0, no net electric charge and a mass slightly larger than that of a proton. With the exception of hydrogen, nuclei of atoms consist of protons and neutrons, which are therefore collectively referred to as nucleons. The number of protons in a nucleus is the atomic number and defines the type of element the atom forms.

PRACTICE QUIZ
Chapter 10. Kinetics of Radioactive Decay

---

1. In nuclear physics, _________ is a situation in which the quantity of a radioactive isotope remains constant because its production rate (due, e.g., to decay of a parent isotope) is equal to its decay rate.

_________ can only occur in a radioactive decay chain if the half-life of the daughter radionuclide B is much shorter than the half-life of the parent radionuclide A. In such a situation, the decay rate of A, and hence the production rate of B, is approximately constant, because the half-life of A is very long compared to the timescales being considered. The quantity of radionuclide B builds up until the number of B atoms decaying per unit time becomes equal to the number being produced per unit time; the quantity of radionuclide B then reaches a constant, equilibrium value.

a. Cirrocumulus cloud
b. Y-intercept
c. Infinite descending chain
d. Secular equilibrium

2. _________(software) is a molecular editor designed for cross-platform use in computational chemistry, molecular modeling, bioinformatics, materials science, and related are. It is extensible through a plugin architecture.

Features

- Molecular builder/editor for Windows, Linux, and Mac OS X.
- All source code is available under the GNU GPL.
- Translations into Chinese, French, German, Italian, Russian, and Spanish.
- Supports multi-threaded rendering and computation.
- Plugin architecture for developers, including rendering, interactive tools, commands, and Python scripts.
- OpenBabel import of files, input generation for multiple computational chemistry packages, crystallography, and biomolecules.

.

a. EMovie
b. Avogadro
c. Extensible Computational Chemistry Environment
d. ICM-Browser

----

3. In nuclear science, the _________ refers to the radioactive decay of different discrete radioactive decay products as a chained series of transformations. Most radioactive elements do not decay directly to a stable state, but rather undergo a series of decays until eventually a stable isotope is reached.

Decay stages are referred to by their relationship to previous or subsequent stages.

a. Decay correct
b. Decay scheme
c. Decay chain
d. Delta ray

4. The _________ is a non-SI unit of radioactivity. It is defined as

1 Ci = 3.7×10^{10} decays per second.

Its continued use is discouraged.

One _________ is roughly the activity of 1 gram of the radium isotope ^{226}Ra, a substance studied by the _________s.

a. Pierre Curie
b. Curie
c. Decay correct
d. Decay scheme

5.

_________ is a chemical element with the symbol Po and atomic number 84, discovered in 1898 by Marie Sklodowska-Curie and Pierre Curie. A rare and highly radioactive element, _________ is chemically similar to bismuth and tellurium, and it occurs in uranium ores. _________ has been studied for possible use in heating spacecraft.

----

a. Potassium
b. Polonium
c. Promethium
d. Protactinium

ANSWER KEY
Chapter 10. Kinetics of Radioactive Decay

---

1. d
2. b
3. c
4. b
5. b

CHAPTER OUTLINE: KEY TERMS, PEOPLE, PLACES, CONCEPTS
Chapter 11
Modes of Radioactive Decay

- Curie
- Radioactive decay
- Nuclide
- Alpha decay
- Coulomb barrier
- Beta decay
- Beta particle
- Excited state
- Kinetic energy
- Decay energy
- Neutron
- Positron
- Positron emission
- Positron emission tomography
- Electron capture
- GAMMA
- Bremsstrahlung
- Gamma ray
- Internal conversion

Chapter 11. Modes of Radioactive Decay

Conversion factor

Neptunium

Nucleosynthesis

Plutonium

Spontaneous fission

Proton emission

Neutron emission

Isotone

Radionuclide

Atomic mass

Atomic weight

Atomic nucleus

Becquerel

Chain reaction

Enrico Fermi

Lise Meitner

Nuclear chain reaction

Curie

The curie is a non-SI unit of radioactivity. It is defined as

$$1\ \text{Ci} = 3.7 \times 10^{10} \text{ decays per second.}$$

Its continued use is discouraged.

One curie is roughly the activity of 1 gram of the radium isotope ^{226}Ra, a substance studied by the Curies.

Radioactive decay

Radioactive decay is the process by which an atomic nucleus of an unstable atom loses energy by emitting ionizing particles (ionizing radiation). The emission is spontaneous, in that the atom decays without any physical interaction with another particle from outside the atom. Usually, radioactive decay happens due to a process confined to the nucleus of the unstable atom, but, on occasion (as with the different processes of electron capture and internal conversion), an inner electron of the radioactive atom is also necessary to the process.

Nuclide

A nuclide is an atomic species characterized by the specific constitution of its nucleus, i.e., by its number of protons Z, its number of neutrons N, and its nuclear energy state.

The word nuclide was proposed by Truman P. Kohman in 1947. Doctor Kohman originally suggested nuclide as referring to a 'species of nucleus' defined by containing a certain number of neutrons and protons. The word thus was originally intended to focus on the nucleus.

Alpha decay

Alpha decay is a type of rioactive decay in which an atomic nucleus emits an alpha particle(two protons and two neutrons) and thereby transforms (or 'decays') into an atom with a mass number 4 less and atomic number 2 less. For example:

238

92U $\rightarrow$ 234

90Th + 4

2He

which can also be written as:

$^{238}U \rightarrow {}^{234}Th + \alpha$,

since an alpha particle is the same as a helium-4 nucleus, which has mass number 4 and atomic number 2. It also has a charge +2, but the charge is usually not written in nuclear equations, which describe nuclear reactions without considering the electrons. This convention is not meant to imply that the nuclei necessarily occur in neutral atoms.

Coulomb barrier

This energy barrier is given by the electrostatic potential energy:

$$U_{coul} = k\frac{q_1 q_2}{r} = \frac{1}{4\pi\epsilon_0}\frac{q_1 q_2}{r}$$

where

k is the Coulomb's constant = 8.9876×10^{9} N m² C^{-2};
ε_0 is the permittivity of free space;
q_1, q_2 are the charges of the interacting particles;
r is the interaction radius.

A positive value of U is due to a repulsive force, so interacting particles are at higher energy levels as they get closer. A negative potential energy indicates a bound state (due to an attractive force).

Coulomb's barrier increases with the atomic numbers (i.e. the number of protons) of the colliding nuclei:

$$U_{coul} = \frac{k\, Z_1\, Z_2\, e^2}{r}$$

where e is the elementary charge, 1.602 176 53×10^{-19} C, and Z_i the corresponding atomic numbers.

To overcome this barrier, nuclei have to collide at high velocities, so their kinetic energies drive them close enough for the strong interaction to take place and bind them together.

According to the kinetic theory of gases, the temperature of a gas is just a measure of the average kinetic energy of the particles in that gas. For normal gases, the Maxwell-Boltzmann distribution gives the fraction of particles moving at a given velocity as a function of gas temperature, and thus the fraction of particles moving at velocities high enough to overcome the Coulomb's barrier can be derived.

In practice, temperatures needed to overcome Coulomb's barrier turn out to be smaller than expected due to quantum-mechanical tunneling, as established by Gamow. The consideration of barrier-penetration through tunneling and the speed distribution gives rise to a limited range of conditions where the fusion can take place, known as the Gamow window.

It was the absence of a Coulomb barrier for the neutron that enabled James Chadwick to discover it.

Beta decay

In nuclear physics, beta decay is a type of radioactive decay in which a beta particle (an electron or a positron) is emitted from an atom. There are two types of beta decay: beta minus and beta plus. In the case of beta decay that produces an electron emission, it is referred to as beta minus, while in the case of a positron emission as beta plus .

Chapter 11. Modes of Radioactive Decay

Beta particle

Beta particles are high-energy, high-speed electrons or positrons emitted by certain types of radioactive nuclei such as potassium-40. The beta particles emitted are a form of ionizing radiation also known as beta rays. The production of beta particles is termed beta decay. They are designated by the Greek letter beta (β).

Excited state

Excitation is an elevation in energy level above an arbitrary baseline energy state. In physics there is a specific technical definition for energy level which is often associated with an atom being excited to an excited state.

In quantum mechanics an excited state of a system (such as an atom, molecule or nucleus) is any quantum state of the system that has a higher energy than the ground state (that is, more energy than the absolute minimum).

Kinetic energy

The kinetic energy of an object is the energy which it possesses due to its motion. It is defined as the work needed to accelerate a body of a given mass from rest to its stated velocity. Having gained this energy during its acceleration, the body maintains this kinetic energy unless its speed changes.

Decay energy

The decay energy is the energy released by a radioactive decay. Radioactive decay is the process in which an unstable atomic nucleus loses energy by emitting ionizing particles and radiation. This decay, or loss of energy, results in an atom of one type, called the parent nuclide transforming to an atom of a different type, called the daughter nuclide.

Neutron

The neutron is a subatomic hadron particle which has the symbol n or n^0, no net electric charge and a mass slightly larger than that of a proton. With the exception of hydrogen, nuclei of atoms consist of protons and neutrons, which are therefore collectively referred to as nucleons. The number of protons in a nucleus is the atomic number and defines the type of element the atom forms.

Positron

The positron is the antiparticle or the antimatter counterpart of the electron. The positron has an electric charge of +1e, a spin of 1/2, and has the same mass as an electron. When a low-energy positron collides with a low-energy electron, annihilation occurs, resulting in the production of two or more gamma ray photons .

Positron emission

Positron emission is converted, via the weak force, to a neutron, releasing a positron (the antimatter counterpart of an electron) and a neutrino.

Isotos which undergo this decay and thereby emit positrons include carbon-11, potassium-40, nitrogen-13, oxygen-15, fluorine-18, and iodine-121. As an example, the following equation describes the beta plus decay of carbon-11 to boron-11, emitting a positron and a neutrino:

Inside protons and neutrons, there are fundamental particles called quarks. The two most common tys of quarks are up quarks, which have a charge of $+{}^{2}{}_{3}$ and down quarks, with a $-{}^{1}{}_{3}$ charge.

Positron emission tomography

Positron emission tomography is a nuclear medicine imaging technique that produces a three-dimensional image or picture of functional processes in the body. The system detects pairs of gamma rays emitted indirectly by a positron-emitting radionuclide (tracer), which is introduced into the body on a biologically active molecule. Three-dimensional images of tracer concentration within the body are then constructed by computer analysis.

Electron capture

Electron capture is a process in which a proton-rich nuclide absorbs an inner atomic electron (changing a nuclear proton to a neutron) and simultaneously emits a neutrino. Various photon emissions follow, in order to allow the energy of the atom to fall to the ground state of the new nuclide.

Electron capture is the primary decay mode for isotopes with a relative superabundance of protons in the nucleus, but with insufficient energy difference between the isotope and its prospective daughter (with one less positive charge) for the nuclide to decay by emitting a positron.

GAMMA

GAMMA experiment is a study of: a) Primary cosmic ray energy spectra and elemental composition (abundances of the elements) at energies 10^{15}-10^{18}eV (so called knee energy region) ; b) Galactic diffuse gamma-ray intensity at energies 10^{14}-10^{15}eV ; c) Extensive Air Showers (EAS) at the mountain level by the ground-based EAS array and underground muon scintillation counters; d) Hard jets production at energies ~10^{16}eV by the muon multi-core shower events.

GAMMA experiment is deployed on the South side of Mount Aragats in Armenia (Cosmic-ray observatory). The facility consists of a ground-based extensive air shower (EAS) array of 33 surface detection stations and 150 underground muon detectors.

Bremsstrahlung — Bremsstrahlung is electromagnetic radiation produced by the deceleration of a charged particle when deflected by another charged particle, typically an electron by an atomic nucleus. The moving particle loses kinetic energy, which is converted into a photon because energy is conserved. The term is also used to refer to the process of producing the radiation.

Gamma ray — Gamma radiation, also known as gamma rays or hyphenated as gamma-rays (especially in astronomy, by analogy with X-rays) and denoted as γ, is electromagnetic radiation of high frequency (very short wavelength). Gamma rays are usually naturally produced on Earth by decay of high energy states in atomic nuclei (gamma decay). An extraterrestrial source of gamma rays originates from atmospheric interaction with cosmic rays.

Internal conversion — Internal conversion is a radioactive decay process where an excited nucleus interacts with an electron in one of the lower atom orbitals, causing the electron to be emitted from the atom. Thus, in an internal conversion process, a high-energy electron is emitted from the radioactive atom, but without beta decay taking place. For this reason, the high-speed electrons from internal conversion are not beta partles (β partles), since the latter come from beta decay.

Conversion factor — A conversion factor changes something to a different version or form. A factor is something that brings results or a cause, while conversion is an action of changing the 'version' of a thing.

In mathematics, specifically algebra, a conversion factor is used to convert a measured quantity to a different unit of measure without changing the relative amount.

Neptunium — Neptunium is a chemical element with the symbol Np and atomic number 93. A radioactive metal, neptunium is the first transuranic element and belongs to the actinide series. Its most stable isotope, ^{237}Np, is a by-product of nuclear reactors and plutonium production and it can be used as a component in neutron detection equipment. Neptunium is also found in trace amounts in uranium ores due to transmutation reactions.

Nucleosynthesis	Nucleosynthesis is the process of creating new atomic nuclei from pre-existing nucleons (protons and neutrons). It is thought that the primordial nucleons themselves were formed from the quark-gluon plasma from the Big Bang as it cooled below two trillion degrees. A few minutes afterward, starting with only protons and neutrons, nuclei up to lithium and beryllium (both with mass number 7) were formed, but only in relatively small amounts.
Plutonium	Plutonium is a transuranic radioactive chemical element with the chemical symbol Pu and atomic number 94. It is an actinide metal of silvery-gray appearance that tarnishes when exposed to air, forming a dull coating when oxidized. The element normally exhibits six allotropes and four oxidation states. It reacts with carbon, halogens, nitrogen and silicon.
Spontaneous fission	Spontaneous fission is a form of radioactive decay characteristic of very heavy isotopes. Because the nuclear binding energy reaches a maximum at a nuclear mass greater than about 60 atomic mass units (u), spontaneous breakdown into smaller nuclei and single particles becomes possible at heavier masses. Because of constraints in constructing the daughter fission product nuclei, spontaneous fission into known nuclides becomes theoretically possible (that is, energetically possible) for many atomic nuclei (nuclide) with a mass greater than or equal to 93 atomic mass units (u), with the possibility increasing as mass number grows above this boundary.
Proton emission	Proton emission is a ty of radioactive decay in which a proton is ejected from a nucleus. Proton emission can occur from high-lying excited states in a nucleus following a beta decay, in which case the process is known as beta-delayed proton emission, or can occur from the ground state (or a low-lying isomer) of very proton-rich nuclei, in which case the process is very similar to alpha decay. For a proton to esca a nucleus, the proton separation energy must be negative - the proton is therefore unbound, and tunnels out of the nucleus in a finite time.
Neutron emission	Neutron emitters to left of lower dashed line

----

Neutron emission is a type of radioactive decay of atoms containing excess neutrons, in which a neutron is simply ejected from the nucleus. Two examples of isotopes which emit neutrons are helium-5 and beryllium-13. However, the decay of helium-5 is also (by definition) a case of alpha-decay.

Many heavy isotopes, most notably californium-252, emit neutrons among the products of a different radioactive decay process, spontaneous fission.

Isotone	Two nuclides are isotones if they have the same neutron number N, but different proton number Z. For example, boron-12 and carbon-13 nuclei both contain 7 neutrons, and so are isotones. Similarly, S-36, Cl-37, Ar-38, K-39, and Ca-40 nuclei are all isotones of 20 because they all contain 20 neutrons. Despite its similarity to the Greek for 'same stretching', the term was formed by replacing the 'p' in 'isotope' with 'n' for 'neutron'.
Radionuclide	A radionuclide is an atom with an unstable nucleus, which is a nucleus characterized by excess energy available to be imparted either to a newly created radiation particle within the nucleus or to an atomic electron. The radionuclide, in this process, undergoes radioactive decay, and emits gamma ray(s) and/or subatomic particles. These particles constitute ionizing radiation.
Atomic mass	The atomic mass is the mass of a specific isotope, most often expressed in unified atomic mass units. The atomic mass is the total mass of protons, neutrons and electrons in a single atom. The atomic mass is sometimes incorrectly used as a synonym of relative atomic mass, average atomic mass and atomic weight; these differ subtly from the atomic mass.
Atomic weight	Atomic weight is a dimensionless physical quantity, the ratio of the average mass of atoms of an element (from a given source) to 1/12 of the mass of an atom of carbon-12 (known as the unified atomic mass unit). The term is usually used, without further qualification, to refer to the standard atomic weights published at regular intervals by the International Union of Pure and Applied Chemistry (IUPAC) and which are intended to be applicable to normal laboratory materials.

Atomic nucleus	The nucleus is the very dense region consisting of protons and neutrons at the center of an atom. It was discovered in 1911, as a result of Ernest Rutherford's interpretation of the famous 1909 Rutherford experiment performed by Hans Geiger and Ernest Marsden, under the direction of Rutherford. The proton-neutron model of nucleus was proposed by Dmitry Ivanenko in 1932. Almost all of the mass of an atom is located in the nucleus, with a very small contribution from the orbiting electrons. The diameter of the nucleus is in the range of 1.75 fm (femtometre) (1.75×10^{-15} m) for hydrogen (the diameter of a single proton) to about 15 fm for the heaviest atoms, such as uranium. These dimensions are much smaller than the diameter of the atom itself (nucleus + electronic cloud), by a factor of about 23,000 (uranium) to about 145,000 (hydrogen). The branch of physics concerned with studying and understanding the atomic nucleus, including its composition and the forces which bind it together, is called nuclear physics.
Becquerel	The becquerel is the SI-derived unit of radioactivity. One Bq is defined as the activity of a quantity of radioactive material in which one nucleus decays per second. The Bq unit is therefore equivalent to s^{-1}. The becquerel is named for Henri Becquerel, who shared a Nobel Prize with Pierre and Marie Curie for their work in discovering radioactivity.
Chain reaction	A chain reaction is a sequence of reactions where a reactive product or by-product causes additional reactions to take place. In a chain reaction, positive feedback leads to a self-amplifying chain of events. Chain reactions are one way in which systems which are in thermodynamic non-equilibrium can release energy or inease entropy in order to reach a state of higher entropy.
Enrico Fermi	Enrico Fermi was an Italian-born, naturalized American physicist particularly known for his work on the development of the first nuclear reactor, Chicago Pile-1, and for his contributions to the development of quantum theory, nuclear and particle physics, and statistical mechanics. He was awarded the 1938 Nobel Prize in Physics for his work on induced radioactivity.

Fermi is widely regarded as one of the leading scientists of the 20th century, highly accomplished in both theory and experiment.

Lise Meitner	Lise Meitner, FRS (7 November 1878 - 27 October 1968) was an Austrian, later Swedish, physicist who worked on radioactivity and nuclear physics. Meitner was part of the team that discovered nuclear fission, an achievement for which her colleague Otto Hahn was awarded the Nobel Prize. Meitner is often mentioned as one of the most glaring examples of women's scientific achievement overlooked by the Nobel committee.
Nuclear chain reaction	A nuclear chain reaction occurs when one nuclear reaction causes an average of one or more nuclear reactions, thus leading to a self-propagating series of these reactions. The specific nuclear reaction may be the fission of heavy isotopes (e.g. ^{235}U) or the fusion of light isotopes (e.g. ^{2}H and ^{3}H). The nuclear chain reaction releases several million times more energy per reaction than any chemical reaction.

----

1. Excitation is an elevation in energy level above an arbitrary baseline energy state. In physics there is a specific technical definition for energy level which is often associated with an atom being excited to an _________.

   In quantum mechanics an _________ of a system (such as an atom, molecule or nucleus) is any quantum state of the system that has a higher energy than the ground state (that is, more energy than the absolute minimum).

   a. Imaginary time
   b. Instrumentalist interpretation
   c. Excited state
   d. Interatomic Coulombic Decay

2. A _________ is an atom with an unstable nucleus, which is a nucleus characterized by excess energy available to be imparted either to a newly created radiation particle within the nucleus or to an atomic electron. The _________, in this process, undergoes radioactive decay, and emits gamma ray(s) and/or subatomic particles. These particles constitute ionizing radiation.

   a. Radiosensitivity
   b. Radionuclide
   c. Sievert
   d. Spinthariscope

3. A _________ occurs when one nuclear reaction causes an average of one or more nuclear reactions, thus leading to a self-propagating series of these reactions. The specific nuclear reaction may be the fission of heavy isotopes (e.g. ^{235}U) or the fusion of light isotopes (e.g. ^{2}H and ^{3}H). The _________ releases several million times more energy per reaction than any chemical reaction.

   a. Nuclear chain reaction
   b. Chemical shift
   c. Chiral model
   d. Cluster decay

4. _________ is a type of rioactive decay in which an atomic nucleus emits an alpha particle(two protons and two neutrons) and thereby transforms (or 'decays') into an atom with a mass number 4 less and atomic number 2 less. For example:

   238

   92U → 234

90Th + 4

2He

which can also be written as:

$^{238}U \rightarrow {}^{234}Th + \alpha$,

since an alpha particle is the same as a helium-4 nucleus, which has mass number 4 and atomic number 2. It also has a charge +2, but the charge is usually not written in nuclear equations, which describe nuclear reactions without considering the electrons. This convention is not meant to imply that the nuclei necessarily occur in neutral atoms.

a. Alpha decay
b. Annihilation radiation
c. Effective dose
d. Ekanite

5. _________ is the process by which an atomic nucleus of an unstable atom loses energy by emitting ionizing particles (ionizing radiation). The emission is spontaneous, in that the atom decays without any physical interaction with another particle from outside the atom. Usually, _________ happens due to a process confined to the nucleus of the unstable atom, but, on occasion (as with the different processes of electron capture and internal conversion), an inner electron of the radioactive atom is also necessary to the process.

a. 5 yen coin
b. Background radiation
c. Background radiation equivalent time
d. Radioactive decay

ANSWER KEY
Chapter 11. Modes of Radioactive Decay

---

1. c
2. b
3. a
4. a
5. d

CHAPTER OUTLINE: KEY TERMS, PEOPLE, PLACES, CONCEPTS

Chapter 12

Production of Radionuclides

- Becquerel
- Cyclotron
- Neutron
- Neutron activation
- Nuclear reactor
- Radionuclide
- Generator
- Actinium
- Neptunium
- Thorium
- Uranium
- Cosmic ray
- Fluence
- Half-life
- Secular equilibrium
- Transient
- Transient equilibrium
- Specific activity
- Neutron capture

Chapter 12. Production of Radionuclides

- Fissile
- Nuclear fission product
- Nuclear fission
- Nuclear transmutation
- Nuclide
- Chain reaction
- Nuclear chain reaction
- Nuclear fusion
- Control rod
- Coolant
- Nuclear power
- Tokamak
- Laser
- Coulomb barrier
- Positron
- Positron emission
- Positron emission tomography
- Nuclear reaction
- Threshold energy

Binding energy

Kinetic energy

Charged particle

Photon

Standing wave

X-rays

Power transmission

Transmission

CHAPTER HIGHLIGHTS: KEY TERMS, PEOPLE, PLACES, CONCEPTS
Chapter 12. Production of Radionuclides

Becquerel

The becquerel is the SI-derived unit of radioactivity. One Bq is defined as the activity of a quantity of radioactive material in which one nucleus decays per second. The Bq unit is therefore equivalent to s^{-1}. The becquerel is named for Henri Becquerel, who shared a Nobel Prize with Pierre and Marie Curie for their work in discovering radioactivity.

Cyclotron

A cyclotron is a compact type of particle accelerator in which charged particles in a static magnetic field are travelling outwards from the center along a spiral path and get accelerated by radio frequency electromagnetic fields.

The cyclotron was invented by Leó Szilárd and first manufactured and patented by Ernest Lawrence, of the University of California, Berkeley who started operating it in 1932 The first European cyclotron was constructed in Leningrad in the physics department of the Radium Institute, headed by Vitali Khlopin.

Neutron

The neutron is a subatomic hadron particle which has the symbol n or n^0, no net electric charge and a mass slightly larger than that of a proton. With the exception of hydrogen, nuclei of atoms consist of protons and neutrons, which are therefore collectively referred to as nucleons. The number of protons in a nucleus is the atomic number and defines the type of element the atom forms.

Neutron activation

Neutron activation is the process in which neutron radiation induces radioactivity in materials, and occurs when atomic nuclei capture free neutrons, becoming heavier and entering excited states. The excited nucleus often decays immediately by emitting particles such as neutrons, protons, or alpha particles. The neutron capture, even after any intermediate decay, often results in the formation of an unstable activation product.

Nuclear reactor

A nuclear reactor is a device to initiate and control a sustained nuclear chain reaction. Most commonly they are used for generating electricity and for the propulsion of ships. Usually heat from nuclear fission is passed to a working fluid (water or gas), which runs through turbines that power either ship's propellers or generators.

Radionuclide

A radionuclide is an atom with an unstable nucleus, which is a nucleus characterized by excess energy available to be imparted either to a newly created radiation particle within the nucleus or to an atomic electron. The radionuclide, in this process, undergoes radioactive decay, and emits gamma ray(s) and/or subatomic particles. These particles constitute ionizing radiation.

Generator

In category theory in mathematics a generator (or separator) of a category C is an object G of the category, such that for any two different morphisms $f, g : X \to Y$ in C, there is a morphism $h : G \to X$, such that the compositions $f \circ h \neq g \circ h$.

Generators are central to the definition of Grothendieck categories.

Examples

- In the category of abelian groups, the group of integers $\mathbf{Z}$ is a generator: If f and g are different, then there is an element $x \in X$, such that $f(x) \neq g(x)$.

Actinium

Actinium is a radioactive chemical element with the symbol Ac and atomic number 89, which was discovered in 1899. It was the first non-primordial radioactive element to be isolated. Polonium, radium and radon were observed before actinium, but they were not isolated until 1902. Actinium gave the name to the actinide series, a group of 15 similar elements between actinium and lawrencium in the periodic table.

A soft, silvery-white radioactive metal, actinium reacts rapidly with oxygen and moisture in air forming a white coating of actinium oxide that prevents further oxidation.

Neptunium

Neptunium is a chemical element with the symbol Np and atomic number 93. A radioactive metal, neptunium is the first transuranic element and belongs to the actinide series. Its most stable isotope, ^{237}Np, is a by-product of nuclear reactors and plutonium production and it can be used as a component in neutron detection equipment. Neptunium is also found in trace amounts in uranium ores due to transmutation reactions.

Thorium

Thorium is a natural radioactive chemical element with the symbol Th and atomic number 90. It was discovered in 1828 and named after Thor, the Norse god of thunder.

In nature, virtually all thorium is found as thorium-232, and it decays by emitting an alpha particle, and has a half-life of about 14.05 billion years (other, trace-level isotopes of thorium are short-lived intermediates of decay chains). It is estimated to be about four times more abundant than uranium in the Earth's crust and is a by-product of the extraction of rare earths from monazite sands.

Uranium

Uranium is a silvery-white metallic chemical element in the actinide series of the periodic table, with atomic number 92. It is assigned the chemical symbol U. A uranium atom has 92 protons and 92 electrons, of which 6 are valence electrons. The uranium nucleus binds between 141 and 146 neutrons, establishing six isotopes (U-233 through U-238), the most common of which are uranium-238 (146 neutrons) and uranium-235 (143 neutrons). All isotopes are unstable and uranium is weakly radioactive.

Cosmic ray

Cosmic rays are energetic charged subatomic particles, originating in outer space. They may produce secondary particles that penetrate the Earth's atmosphere and surface. The term ray is historical as cosmic rays were thought to be electromagnetic radiation.

Fluence

In physics, fluence is the flux (either particle or radiative flux) integrated over time. For particles, it is defined as the total number of particles that intersect a unit area in a specific time interval of interest, and has units of m^{-2} (number of particles per meter squared). Fluence can also be used to describe the energy delivered per unit area, in which case it has units of J/m^2.

Half-life

Half-life, abbreviated $t_{1/2}$, is the period of time it takes for the amount of a substance undergoing decay to decrease by half. The name was originally used to describe a characteristic of unstable atoms (radioactive decay), but it may apply to any quantity which follows a set-rate decay.

	The original term, dating to 1907, was 'half-life period', which was later shortened to 'half-life' in the early 1950s.
Secular equilibrium	In nuclear physics, Secular equilibrium is a situation in which the quantity of a radioactive isotope remains constant because its production rate (due, e.g., to decay of a parent isotope) is equal to its decay rate. Secular equilibrium can only occur in a radioactive decay chain if the half-life of the daughter radionuclide B is much shorter than the half-life of the parent radionuclide A. In such a situation, the decay rate of A, and hence the production rate of B, is approximately constant, because the half-life of A is very long compared to the timescales being considered. The quantity of radionuclide B builds up until the number of B atoms decaying per unit time becomes equal to the number being produced per unit time; the quantity of radionuclide B then reaches a constant, equilibrium value.
Transient	A transient event is a short-lived burst of energy in a system caused by a sudden change of state. The source of the transient energy may be an internal event or a nearby event. The energy then couples other parts of the system, typically appearing as a short burst of oscillation.
Transient equilibrium	In nuclear physics, transient equilibrium is a situation in which equilibrium is reached by a parent-daughr radioactive isotope pair where the half-life of the daughr is shorr than the half-life of the parent. Contrary to Secular equilibrium, the half-life of the daughr is not negligible compare to parent's half-life. An example of this is the Molybdenum 99 generator producing chnetium 99 for nuclear medicine diagnostic procedures.
Specific activity	In nuclear sciences and technologies, 'activity' is the SI quantity related to the phenomenon of natural and artificial radioactivity. The SI unit of 'activity' is becquerel, Bq, while that of 'specific activity' is Bq/kg. The old unit of 'activity' was curie, Ci, while that of 'specific activity' was Ci/g.

Chapter 12. Production of Radionuclides

Neutron capture	Neutron capture is a kind of nuclear reaction in which an atomic nucleus collides with one or more neutrons and they merge to form a heavier nucleus. Sie neutrons have no electric charge they can enter a nucleus more easily than positively charged protons, which are repelled electrostatically. Neutron capture plays an important role in the cosmic nucleosynthesis of heavy elements.
Fissile	In nuclear engineering, a fissile material is one that is capable of sustaining a chain reaction of nuclear fission. By definition, fissile materials can sustain a chain reaction with neutrons of any energy. The predominant neutron energy may be typified by either slow neutrons (i.e. a thermal system) or fast neutrons.
Nuclear fission product	Nuclear fission products are the atomic fragments left after a large atomic nucleus fissions. Typically, a large nucleus like that of uranium fissions by splitting into two smaller nuclei, along with a few neutrons and release of energy in the form of heat (kinetic energy of the nuclei) and gamma rays. The two smaller nuclei are the 'fission products'.
Nuclear fission	In nuclear physics and nuclear chemistry, nuclear fission refers to either a nuclear reaction or a radioactive decay process in which the nucleus of an atom splits into smaller parts (lighter nuclei), often producing free neutrons and photons (in the form of gamma rays), and releasing a very large amount of energy, even by the energetic standards of radioactive decay. The two nuclei produced are most often of comparable but slightly different sizes, typically with a mass ratio of products of about 3 to 2, for common fissile isotopes. Most fissions are binary fissions (producing two charged fragments), but occasionally (2 to 4 times per 1000 events), three positively charged fragments are produced, in a ternary fission.
Nuclear transmutation	Nuclear transmutation is the conversion of one chemical eleme or isotope io another. In other words, atoms of one eleme can be changed io atoms of other eleme by 'transmutation'. This occurs either through nuclear reactions (in which an outside particle reacts with a nucleus), or through radioactive decay (where no outside particle is needed).
Nuclide	A nuclide is an atomic species characterized by the specific constitution of its nucleus, i.e., by its number of protons Z, its number of neutrons N, and its nuclear energy state.

The word nuclide was proposed by Truman P. Kohman in 1947. Doctor Kohman originally suggested nuclide as referring to a 'species of nucleus' defined by containing a certain number of neutrons and protons. The word thus was originally intended to focus on the nucleus.

Chain reaction

A chain reaction is a sequence of reactions where a reactive product or by-product causes additional reactions to take place. In a chain reaction, positive feedback leads to a self-amplifying chain of events.

Chain reactions are one way in which systems which are in thermodynamic non-equilibrium can release energy or inease entropy in order to reach a state of higher entropy.

Nuclear chain reaction

A nuclear chain reaction occurs when one nuclear reaction causes an average of one or more nuclear reactions, thus leading to a self-propagating series of these reactions. The specific nuclear reaction may be the fission of heavy isotopes (e.g. ^{235}U) or the fusion of light isotopes (e.g. ^{2}H and ^{3}H). The nuclear chain reaction releases several million times more energy per reaction than any chemical reaction.

Nuclear fusion

Nuclear fusion is the process by which two or more atomic nuclei join together, or 'fuse', to form a single heavier nucleus. This is usually accompanied by the release of large quantities of energy. Fusion is the process that powers active stars, the hydrogen bomb and some experimental devices examining fusion power for electrical generation.

Control rod

A control rod is a rod made of chemical elements capable of absorbing many neutrons without fissioning themselves. They are used in nuclear reactors to control the rate of fission of uranium and plutonium. Because these elements have different capture cross sections for neutrons of varying energies, the compositions of the control rods must be designed for the neutron spectrum of the reactor it is supposed to control. Light water reactors (BWR, PWR) and heavy water reactors (HWR) operate with 'thermal' neutrons, whereas breeder reactors operate with 'fast' neutrons.

Coolant

A coolant is a fluid which flows through a device to prevent its overheating, transferring the heat produced by the device to other devices that use or dissipate it. An ideal coolant has high thermal capacity, low viscosity, is low-cost, non-toxic, and chemically inert, neither causing nor promoting corrosion of the cooling system. Some applications also require the coolant to be an electrical insulator.

Chapter 12. Production of Radionuclides

Nuclear power

Nuclear power is the use of sustained nuclear fission to generate heat and electricity. Nuclear power plants provide about 6% of the world's energy and 13-14% of the world's electricity, with the U.S., France, and Japan together accounting for about 50% of nuclear generated electricity. In 2007, the IAEA reported there were 439 nuclear power reactors in operation in the world, operating in 31 countries.

Tokamak

A tokamak is a device using a magnetic field to confine a plasma in the shape of a torus (doughnut). Achieving a stable plasma equilibrium requires magnetic field lines that move around the torus in a helical shape. Such a helical field can be generated by adding a toroidal field (traveling around the torus in circles) and a poloidal field (traveling in circles orthogonal to the toroidal field).

Laser

A laser is a device that emits light (electromagnetic radiation) through a process of optical amplification based on the stimulated emission of photons. The term 'laser' originated as an acronym for Light Amplification by Stimulated Emission of Radiation. The emitted laser light is notable for its high degree of spatial and temporal coherence, unattainable using other technologies.

Coulomb barrier

This energy barrier is given by the electrostatic potential energy:

$$U_{coul} = k\frac{q_1 q_2}{r} = \frac{1}{4\pi\epsilon_0}\frac{q_1 q_2}{r}$$

where

k is the Coulomb's constant = 8.9876×10^{9} N m² C^{-2};
ε_0 is the permittivity of free space;
q_1, q_2 are the charges of the interacting particles;
r is the interaction radius.

A positive value of U is due to a repulsive force, so interacting particles are at higher energy levels as they get closer. A negative potential energy indicates a bound state (due to an attractive force).

Coulomb's barrier increases with the atomic numbers (i.e. the number of protons) of the colliding nuclei:

$$U_{coul} = \frac{k\, Z_1\, Z_2\, e^2}{r}$$

where e is the elementary charge, 1.602 176 53×10^{-19} C, and Z_i the corresponding atomic numbers.

To overcome this barrier, nuclei have to collide at high velocities, so their kinetic energies drive them close enough for the strong interaction to take place and bind them together.

According to the kinetic theory of gases, the temperature of a gas is just a measure of the average kinetic energy of the particles in that gas. For normal gases, the Maxwell-Boltzmann distribution gives the fraction of particles moving at a given velocity as a function of gas temperature, and thus the fraction of particles moving at velocities high enough to overcome the Coulomb's barrier can be derived.

In practice, temperatures needed to overcome Coulomb's barrier turn out to be smaller than expected due to quantum-mechanical tunneling, as established by Gamow. The consideration of barrier-penetration through tunneling and the speed distribution gives rise to a limited range of conditions where the fusion can take place, known as the Gamow window.

It was the absence of a Coulomb barrier for the neutron that enabled James Chadwick to discover it.

Positron

The positron is the antiparticle or the antimatter counterpart of the electron. The positron has an electric charge of +1e, a spin of 1/2, and has the same mass as an electron. When a low-energy positron collides with a low-energy electron, annihilation occurs, resulting in the production of two or more gamma ray photons .

Chapter 12. Production of Radionuclides

Positron emission

Positron emission is converted, via the weak force, to a neutron, releasing a positron (the antimatter counterpart of an electron) and a neutrino.

Isotos which undergo this decay and thereby emit positrons include carbon-11, potassium-40, nitrogen-13, oxygen-15, fluorine-18, and iodine-121. As an example, the following equation describes the beta plus decay of carbon-11 to boron-11, emitting a positron and a neutrino:

Inside protons and neutrons, there are fundamental particles called quarks. The two most common tys of quarks are up quarks, which have a charge of $+{}^{2}{}_{3}$ and down quarks, with a $-{}^{1}{}_{3}$ charge.

Positron emission tomography

Positron emission tomography is a nuclear medicine imaging technique that produces a three-dimensional image or picture of functional processes in the body. The system detects pairs of gamma rays emitted indirectly by a positron-emitting radionuclide (tracer), which is introduced into the body on a biologically active molecule. Three-dimensional images of tracer concentration within the body are then constructed by computer analysis.

Nuclear reaction

In nuclear physics and nuclear chemistry, a nuclear reaction is semantically considered to be the process in which two nuclei, or else a nucleus of an atom and a subatomic particle (such as a proton, neutron, or high energy electron) from outside the atom, collide to produce products different from the initial particles. In principle, a reaction can involve more than three particles colliding, but because the probability of three or more nuclei to meet at the same time at the same place is much less than for two nuclei, such an event is exceptionally rare. Radioactive decays can be considered to be spontaneous nuclear reactions, in as much as there is such a thing as a spontaneous chemical reaction.

Threshold energy

In particle physics, the threshold energy for production of a particle is the minimum kinetic energy a pair of traveling particles must have when they collide. The threshold energy is always grear than or equal to the rest energy of the desired particle. In most cases, since momentum is also conserved, the threshold energy is significantly grear than the rest energy of the desired particle - and thus there will still be considerable kinetic energy in the final particles.

Binding energy

Binding energy is the mechanical energy required to disassemble a whole into separate parts. A bound system typically has a lower potential energy than its constituent parts; this is what keeps the system together--often this means that energy is released upon the creation of a bound state. The usual convention is that this corresponds to a positive binding energy.

Kinetic energy

The kinetic energy of an object is the energy which it possesses due to its motion. It is defined as the work needed to accelerate a body of a given mass from rest to its stated velocity. Having gained this energy during its acceleration, the body maintains this kinetic energy unless its speed changes.

Charged particle

In physics, a charged particle is a particle with an electric charge. It may be either a subatomic particle or an ion. A collection of charged particles, or even a gas containing a proportion of charged particles, is called a plasma, which is called the fourth state of matter because its properties are quite different from solids, liquids and gases (plasma is the most common state of matter in the universe).

Photon

In physics, a photon is an elementary particle, the quantum of light and all other forms of electromagnetic radiation, and the force carrier for the electromagnetic force. The effects of this force are easily observable at both the microscopic and macroscopic level, because the photon has no rest mass; this allows for interactions at long distances. Like all elementary particles, photons are currently best explained by quantum mechanics and exhibit wave-particle duality, exhibiting properties of both waves and particles.

Standing wave

In physics, a standing wave - also known as a stationary wave - is a wave that remains in a constant position.

This phenomenon can occur because the medium is moving in the opposite direction to the wave, or it can arise in a stationary medium as a result of interference between two waves traveling in opposite directions. In the second case, for waves of equal amplitude traveling in opposing directions, there is on average no net propagation of energy.

X-rays

X-radiation (composed of X-rays) is a form of electromagnetic radiation. X-rays have a wavelength in the range of 10 to 0.01 nanometers, corresponding to frequencies in the range 30 petahertz to 30 exahertz (3×10^{16} Hz to 3×10^{19} Hz) and energies in the range 120 eV to 120 keV. They are shorter in wavelength than UV rays. In many languages, X-radiation is called Röntgen radiation after Wilhelm Conrad Röntgen, who is generally credited as their discoverer, and who had called them X-rays to signify an unknown type of radiation.[:1-2]

X-rays from about 0.12 to 12 keV (10 to 0.10 nm wavelength), are classified as soft X-rays, and from about 12 to 120 keV (0.10 to 0.010 nm wavelength) as hard X-rays, due to their penetrating abilities.

Power transmission	Power transmission is the movement of energy from its place of generation to a location where it is applied to performing useful work. Power is defined formally as units of energy per unit time. In SI units: $$\text{watt} = \frac{\text{joule}}{\text{second}} = \frac{\text{newton} \times \text{meter}}{\text{second}}$$ Since the development of technology, transmission and storage systems have been of immense interest to technologists and technology users.
Transmission	A machine consists of a power source and a power transmission system, which provides controlled application of the power. Merriam-Webster defines transmission as: an assembly of parts including the speed-changing gears and the propeller shaft by which the power is transmitted from an engine to a live axle. Often transmission refers simply to the gearbox that uses gears and gear trains to provide speed and torque conversions from a rotating power source to another device.

----

1. _________ is the process in which neutron radiation induces radioactivity in materials, and occurs when atomic nuclei capture free neutrons, becoming heavier and entering excited states. The excited nucleus often decays immediately by emitting particles such as neutrons, protons, or alpha particles. The neutron capture, even after any intermediate decay, often results in the formation of an unstable activation product.

   a. Neutron-induced swelling
   b. Radiation damage
   c. Radiation effect
   d. Neutron activation

2. The _________ is the SI-derived unit of radioactivity. One Bq is defined as the activity of a quantity of radioactive material in which one nucleus decays per second. The Bq unit is therefore equivalent to s^{-1}. The _________ is named for Henri _________, who shared a Nobel Prize with Pierre and Marie Curie for their work in discovering radioactivity.

   a. Revolutions per minute
   b. Cirrocumulus cloud
   c. Becquerel
   d. 1,8-Diaminonaphthalene

3. A _________ is an atom with an unstable nucleus, which is a nucleus characterized by excess energy available to be imparted either to a newly created radiation particle within the nucleus or to an atomic electron. The _________, in this process, undergoes radioactive decay, and emits gamma ray(s) and/or subatomic particles. These particles constitute ionizing radiation.

   a. Radiosensitivity
   b. Rainout
   c. Radionuclide
   d. Spinthariscope

4. _________ is a silvery-white metallic chemical element in the actinide series of the periodic table, with atomic number 92. It is assigned the chemical symbol U. A _________ atom has 92 protons and 92 electrons, of which 6 are valence electrons. The _________ nucleus binds between 141 and 146 neutrons, establishing six isotopes (U-233 through U-238), the most common of which are _________-238 (146 neutrons) and _________-235 (143 neutrons). All isotopes are unstable and _________ is weakly radioactive.

----

a. Eisenstein ideal
b. Uranium
c. Titanium
d. Transfermium element

5. A _________ is a compact type of particle accelerator in which charged particles in a static magnetic field are travelling outwards from the center along a spiral path and get accelerated by radio frequency electromagnetic fields.

The _________ was invented by Leó Szilárd and first manufactured and patented by Ernest Lawrence, of the University of California, Berkeley who started operating it in 1932 The first European _________ was constructed in Leningrad in the physics department of the Radium Institute, headed by Vitali Khlopin.

a. Cyclotron resonance
b. Dielectric wall accelerator
c. Cyclotron
d. Dynamitron

ANSWER KEY
Chapter 12. Production of Radionuclides

---

1. d
2. c
3. c
4. b
5. c

CHAPTER OUTLINE: KEY TERMS, PEOPLE, PLACES, CONCEPTS

Chapter 13

Waveguide Theory

______ S band

______ X band

______ Ampere

______ Magnetism

______ Dirichlet boundary condition

______ Neumann boundary condition

______ Wave equation

______ Particle accelerator

______ Bessel function

______ Particle acceleration

______ Acceleration

______ Dispersion

______ Group velocity

______ Phase velocity

______ Wavelength

______ Doppler effect

______ Relativistic Doppler effect

______ Reflection

______ Inverse

CHAPTER HIGHLIGHTS: KEY TERMS, PEOPLE, PLACES, CONCEPTS
Chapter 13. Waveguide Theory

S band

The S band is part of the microwave band of the electromagnetic spectrum. It is defined by an IEEE standard for radio waves with frequencies that range from 2 to 4 GHz, crossing the conventional boundary between UHF and SHF at 3.0 GHz. The S band is used by weather radar, surface ship radar, and some communications satellites, especially those used by NASA to communicate with the Space Shuttle and the International Space Station.

X band

The X band is a segment of the microwave radio region of the electromagnetic spectrum. In some cases, such as in communication engineering, the frequency range of X band is rather indefinitely set at approximately 7.0 to 11.2 gigahertz (GHz). In radar engineering, the frequency range is specified by the IEEE at 8.0 to 12.0 GHz.

Ampere

The ampere is the SI unit of electric current (symbol: I) and is one of the seven SI base units. It is named after André-Marie Ampère (1775-1836), French mathematician and physicist, considered the father of electrodynamics. In practice, its name is often shortened to amp.

Magnetism

Magnetism is a property of materials that respond to an applied magnetic field. Permanent magnets have persistent magnetic fields caused by ferromagnetism. That is the strongest and most familiar type of magnetism.

Dirichlet boundary condition

In mathematics, the Dirichlet boundary condition is a type of boundary condition. When imposed on an ordinary or a partial differential equation, it specifies the values a solution needs to take on the boundary of the domain. The question of finding solutions to such equations is known as the Dirichlet problem.

- For an ordinary differential equation, for instance:

$$y'' + y = 0$$

the Dirichlet boundary conditions on the interval $[a, b]$ take the form:

$$y(a) = \alpha \text{ and } y(b) = \beta$$

where α and β are given numbers.

Neumann boundary condition	In mathematics, the Neumann boundary condition is a type of boundary condition. When imposed on an ordinary or a partial differential equation, it specifies the values that the derivative of a solution is to take on the boundary of the domain. • For an ordinary differential equation, for instance: $y'' + y = 0$ the Neumann boundary conditions on the interval $[a, b]$ take the form: $y'(a) = \alpha \text{ and } y'(b) = \beta$ where α and β are given numbers.
Wave equation	The wave equation is an important second-order linear partial differential equation for the description of waves - as they occur in physics - such as sound waves, light waves and water waves. It arises in fields like acoustics, electromagnetics, and fluid dynamics. Historically, the problem of a vibrating string such as that of a musical instrument was studied by Jean le Rond d'Alembert, Leonhard Euler, Daniel Bernoulli, and Joseph-Louis Lagrange.
Particle accelerator	A particle accelerator is a device that uses electromagnetic fields to propel charged particles to high speeds and to contain them in well-defined beams. The two basic classes of accelerators are Alongside their best known use in particle physics as colliders (e.g. LHC, RHIC, Tevatron), particle accelerators are used in a large variety of applications, including particle therapy for oncological purposes, and as synchrotron light sources for e.g. condensed matter physics. Because colliders can give evidence on the structure of the subatomic world, accelerators were commonly referred to as atom smashers in the 20th century.

Bessel function

In mathematics, Bessel functions, first defined by the mathematician Daniel Bernoulli and generalized by Friedrich Bessel, are canonical solutions y(x) of Bessel's differential equation:

$$x^2\frac{d^2y}{dx^2}+x\frac{dy}{dx}+(x^2-\alpha^2)y=0$$

for an arbitrary real or complex number α (the order of the Bessel function); the most common and important cases are for α an integer or half-integer.

Although α and −α produce the same differential equation, it is conventional to define different Bessel functions for these two orders (e.g., so that the Bessel functions are mostly smooth functions of α). Bessel functions are also known as cylinder functions or cylindrical harmonics because they are found in the solution to Laplace's equation in cylindrical coordinates.

Particle acceleration

In a compressible sound transmission medium - mainly air - air particles get an accelerated motion: the particle acceleration or sound acceleration with the symbol a in metre/second^2. In acoustics or physics, acceleration (symbol: a) is defined as the rate of change (or time derivative) of velocity. It is thus a vector quantity with dimension lengthtime^2.

Acceleration

In physics, acceleration is the rate of change of velocity with time. In one dimension, acceleration is the rate at which something speeds up or slows down. However, since velocity is a vector, acceleration describes the rate of change of both the magnitude and the direction of velocity.

Dispersion

In optics, dispersion is the phenomenon in which the phase velocity of a wave depends on its frequency, or alternatively when the group velocity depends on the frequency. Media having such a property are termed dispersive media. Dispersion is sometimes called chromatic dispersion to emphasize its wavelength-dependent nature, or group-velocity dispersion (GVD) to emphasize the role of the group velocity.

Group velocity

The group velocity of a wave is the velocity with which the overall shape of the wave's amplitudes -- known as the modulation or envelope of the wave -- propagates through space.

For example, imagine what happens if a stone is thrown into the middle of a very still pond. When the stone hits the surface of the water, a circular pattern of waves appears.

Phase velocity

The phase velocity of a wave is the rate at which the phase of the wave propagates in space. This is the speed at which the phase of any one frequency component of the wave travels. For such a component, any given phase of the wave (for example, the crest) will appear to travel at the phase velocity.

Wavelength

In physics, the wavelength of a sinusoidal wave is the spatial period of the wave--the distance over which the wave's shape repeats. It is usually determined by considering the distance between consecutive corresponding points of the same phase, such as crests, troughs, or zero crossings, and is a characteristic of both traveling waves and standing waves, as well as other spatial wave patterns. Wavelength is commonly designated by the Greek letter lambda (λ).

Doppler effect

The Doppler effect is the change in frequency of a wave for an observer moving relative to the source of the wave. It is commonly heard when a vehicle sounding a siren or horn approaches, passes, and reces from an observer. The received frequency is higher (compared to the emitted frequency) during the approach, it is intical at the instant of passing by, and it is lower during the recession.

Relativistic Doppler effect

The relativistic Doppler effect is the change in frequency (and wavelength) of light, caused by the relative motion of the source and the observer (as in the classical Doppler effect), when taking into account effects described by the special theory of relativity.

The relativistic Doppler effect is different from the non-relativistic Doppler effect as the equations include the time dilation effect of special relativity and do not involve the medium of propagation as a reference point. They describe the total difference in observed frequencies and possess the required Lorentz symmetry.

Reflection

Reflection is the change in direction of a wavefront at an interface between two different media so that the wavefront returns into the medium from which it originated. Common examples include the reflection of light, sound and water waves. The law of reflection says that for specular reflection the angle at which the wave is incident on the surface equals the angle at which it is reflected.

Inverse	In many contexts in mathematics the term inverse indicates the opposite of something. This word and its derivatives are used widely in mathematics, as illustrated below. • Inverse element of an element x with respect to a binary operation * with identity element e is an element y such that x * y = y * x = e.

----

1. In physics, the _________ of a sinusoidal wave is the spatial period of the wave--the distance over which the wave's shape repeats. It is usually determined by considering the distance between consecutive corresponding points of the same phase, such as crests, troughs, or zero crossings, and is a characteristic of both traveling waves and standing waves, as well as other spatial wave patterns. _________ is commonly designated by the Greek letter lambda (λ).

   a. Whispering-gallery wave
   b. Quasiperiodicity
   c. Rayleigh wave
   d. Wavelength

2. The _________ is part of the microwave band of the electromagnetic spectrum. It is defined by an IEEE standard for radio waves with frequencies that range from 2 to 4 GHz, crossing the conventional boundary between UHF and SHF at 3.0 GHz. The _________ is used by weather radar, surface ship radar, and some communications satellites, especially those used by NASA to communicate with the Space Shuttle and the International Space Station.

   a. Cirrocumulus cloud
   b. S band
   c. 1,8-Diaminonaphthalene
   d. 1010

3. The _________ is the change in frequency (and wavelength) of light, caused by the relative motion of the source and the observer (as in the classical Doppler effect), when taking into account effects described by the special theory of relativity.

   The _________ is different from the non-_________ as the equations include the time dilation effect of special relativity and do not involve the medium of propagation as a reference point. They describe the total difference in observed frequencies and possess the required Lorentz symmetry.

   a. Cirrocumulus cloud
   b. Flutter
   c. Relativistic Doppler effect
   d. Frequency of optimum transmission

----

4. The _________ is an important second-order linear partial differential equation for the description of waves - as they occur in physics - such as sound waves, light waves and water waves. It arises in fields like acoustics, electromagnetics, and fluid dynamics. Historically, the problem of a vibrating string such as that of a musical instrument was studied by Jean le Rond d'Alembert, Leonhard Euler, Daniel Bernoulli, and Joseph-Louis Lagrange.

   a. Wave equation
   b. Wave impedance
   c. Wave packet
   d. Wave propagation

5. In optics, dispersion is the phenomenon in which the phase velocity of a wave depends on its frequency, or alternatively when the group velocity depends on the frequency. Media having such a property are termed dispersive media. Dispersion is sometimes called chromatic dispersion to emphasize its wavelength-dependent nature, or group-velocity _________(GVD) to emphasize the role of the group velocity.

   a. Distinctness of image
   b. Distortion
   c. Dispersion
   d. F-number

ANSWER KEY
Chapter 13. Waveguide Theory

1. d
2. b
3. c
4. a
5. c

CHAPTER OUTLINE: KEY TERMS, PEOPLE, PLACES, CONCEPTS
Chapter 14
Particle Accelerators in Medicine

- ________ Electrostatic
- ________ Particle accelerator
- ________ X-ray astronomy
- ________ Crystallography
- ________ Diffraction
- ________ Spectroscopy
- ________ Bremsstrahlung
- ________ Synchrotron
- ________ Synchrotron radiation
- ________ Crookes tube
- ________ High voltage
- ________ Positron
- ________ Positron emission
- ________ Positron emission tomography
- ________ Voltage
- ________ Anode
- ________ Cold cathode
- ________ Hot cathode
- ________ Cathode

Chapter 14. Particle Accelerators in Medicine

X-ray tube

Cathode ray

Thermionic emission

William Duane

Work function

Carbon nanotube

Betatron

Cyclotron

Flattening

Microtron

Annihilation

Photon

Storage ring

Synchrotron light source

Monitoring

Dual

Multileaf collimator

Collimator

Ionization

Ionization chamber

Transmission

Laser

Duty cycle

Alpha decay

GAMMA

Internal conversion

Neutron

Neutron emission

Proton emission

Carl David Anderson

Avogadro

Becquerel

William Henry Bragg

William Lawrence Bragg

Curie

Albert Einstein

Enrico Fermi

James Franck

Chapter 14. Particle Accelerators in Medicine

Hartree

Douglas Hartree

Werner Heisenberg

Hertz

Gustav Ludwig Hertz

Max von Laue

Peter Mansfield

Peter Kramer

James Clerk Maxwell

Lise Meitner

Wolfgang Pauli

Planck

Rayleigh

Owen Willans Richardson

Electrostatic	Electrostatics is the branch of physics that deals with the phenomena and properties of stationary or slow-moving (without acceleration) electric charges. Since classical antiquity, it was known that some materials such as amber attract lightweight particles after rubbing. The Greek word for amber, ?λεκτρον electron, was the source of the word 'electricity'.
Particle accelerator	A particle accelerator is a device that uses electromagnetic fields to propel charged particles to high speeds and to contain them in well-defined beams. The two basic classes of accelerators are Alongside their best known use in particle physics as colliders (e.g. LHC, RHIC, Tevatron), particle accelerators are used in a large variety of applications, including particle therapy for oncological purposes, and as synchrotron light sources for e.g. condensed matter physics. Because colliders can give evidence on the structure of the subatomic world, accelerators were commonly referred to as atom smashers in the 20th century.
X-ray astronomy	X-ray astronomy is an observational branch of astronomy which deals with the study of X-ray observation and detection from astronomical objects. X-radiation is absorbed by the Earth's atmosphere, so instruments to detect X-rays must be taken to high altitude by balloons, sounding rockets, and satellites. X-ray astronomy is part of space science.
Crystallography	Crystallography is the science of the arrangement of atoms in solids. The word 'crystallography' derives from the Greek words crystallon = cold drop / frozen drop, with its meaning extending to all solids with some degree of transparency, and grapho = write.

Chapter 14. Particle Accelerators in Medicine

Diffraction	Diffraction refers to various phenomena which occur when a wave encounters an obstacle. Italian scientist Francesco Maria Grimaldi coined the word 'diffraction' and was the first to record accurate observations of the phenomenon in 1665. In classical physics, the diffraction phenomenon is described as the apparent bending of waves around small obstacles and the spreading out of waves past small openings. Similar effects occur when light waves travel through a medium with a varying refractive index or a sound wave through one with varying acoustic impedance.
Spectroscopy	Spectroscopy is the study of the interaction between matter and radiated energy. Historically, spectroscopy originated through the study of visible light dispersed according to its wavelength, e.g., by a prism. Later the concept was expanded greatly to comprise any interaction with radiative energy as a function of its wavelength or frequency.
Bremsstrahlung	Bremsstrahlung is electromagnetic radiation produced by the deceleration of a charged particle when deflected by another charged particle, typically an electron by an atomic nucleus. The moving particle loses kinetic energy, which is converted into a photon because energy is conserved. The term is also used to refer to the process of producing the radiation.
Synchrotron	A synchrotron is a particular type of cyclic particle accelerator originating from the cyclotron in which the guiding magnetic field (bending the particles into a closed path) is time-dependent, being synchronized to a particle beam of increasing kinetic energy. The synchrotron is one of the first accelerator concepts that enable the construction of large-scale facilities, since bending, beam focusing and acceleration can be separated into different components. The first electron synchrotron was constructed by Edwin McMillan in 1945, although the principle had already been published (unknown to him) in a Russian journal by Vladimir Veksler.
Synchrotron radiation	The electromagnetic radiation emitted when charged particles are accelerated radially () is called synchrotron radiation. It is produced in synchrotrons using bending magnets, undulators and/or wigglers. It is similar to cyclotron radiation except that cyclotron radiation is generated by the acceleration of ultrarelativistic charged particles through magnetic fields.

Crookes tube

A Crookes tube is an early experimental elerical discharge tube, invented by English physicist William Crookes and others around 1869-1875, in which cathode rays, streams of elerons, were discovered.

A development of the Geissler tube, it consists of a partially evacuated glass container of various shapes, with two metal elerodes, one at either end. When a high voltage is applied between the elerodes, cathode rays (elerons) travel in straight lines from the cathode to the anode.

High voltage

The term high voltage characterizes electrical circuits in which the voltage used is the cause of particular safety concerns and insulation requirements. High voltage is used in electrical power distribution, in cathode ray tubes, to generate X-rays and particle beams, to demonstrate arcing, for ignition, in photomultiplier tubes, and in high power amplifier vacuum tubes and other industrial and scientific applications.

Definition
The numerical definition of high voltage depends on the context of the discussion.

Positron

The positron is the antiparticle or the antimatter counterpart of the electron. The positron has an electric charge of +1e, a spin of 1/2, and has the same mass as an electron. When a low-energy positron collides with a low-energy electron, annihilation occurs, resulting in the production of two or more gamma ray photons .

Positron emission

Positron emission is converted, via the weak force, to a neutron, releasing a positron (the antimatter counterpart of an electron) and a neutrino.

Isotos which undergo this decay and thereby emit positrons include carbon-11, potassium-40, nitrogen-13, oxygen-15, fluorine-18, and iodine-121. As an example, the following equation describes the beta plus decay of carbon-11 to boron-11, emitting a positron and a neutrino:

	Inside protons and neutrons, there are fundamental particles called quarks. The two most common tys of quarks are up quarks, which have a charge of $+{}^{2}{}_{3}$ and down quarks, with a $-{}^{1}{}_{3}$ charge.
Positron emission tomography	Positron emission tomography is a nuclear medicine imaging technique that produces a three-dimensional image or picture of functional processes in the body. The system detects pairs of gamma rays emitted indirectly by a positron-emitting radionuclide (tracer), which is introduced into the body on a biologically active molecule. Three-dimensional images of tracer concentration within the body are then constructed by computer analysis.
Voltage	Voltage, otherwise known as electrical potential difference or electric tension (denoted ΔV and measured in volts, or joules per coulomb) is the potential difference between two points -- or the difference in electric potential energy per unit charge between two points. Voltage is equal to the work which would have to be done, per unit charge, against a static electric field to move the charge between two points. A voltage may represent either a source of energy (electromotive force), or it may represent lost or stored energy (potential drop).
Anode	An anode is an electrode through which electric current flows into a polarized electrical device. The direction of electric current is, by convention, opposite to the direction of electron flow. In other words, the electrons flow from the anode into, for example, an electrical circuit.
Cold cathode	A cold cathode is a cathode used within nixie tubes, gas discharge lamps, discharge tubes, and some types of vacuum tube which is not electrically heated by the circuit to which it is connected. Such a cathode may be heated to a high temperature in normal circuit operation. Cold-cathode devices A cold-cathode vacuum tube does not rely on external heating of an electrode to provide thermionic emission of electrons.
Hot cathode	In vacuum tubes, a hot cathode is a cathode electrode which emits electrons due to thermionic emission. In the accelerator physics (particle accelerator) community, these are referred to as thermionic cathodes. The heating element is usually an electrical filament.

Cathode	A cathode is an electrode through which electric current flows out of a polarized electrical device. The direction of electric current is, by convention, opposite to the direction of electron flow. Therefore the electrons flow into the polarized electrical device and out of, for example, the connected electrical circuit.
X-ray tube	An X-ray tube is a vacuum tube that produces X-rays. They are used in X-ray machines. X-rays are part of the electromagnetic spectrum, an ionizing radiation with wavelengths shorter than ultraviolet light.
Cathode ray	Cathode rays (also called an electron beam or e-beam) are streams of electrons observed in vacuum tubes. If an evacuated glass tube is equipped with two electrodes and a voltage is applied, the glass opposite of the negative electrode is observed to glow, due to electrons emitted from and travelling perpendicular to the cathode (the electrode connected to the negative terminal of the voltage supply). They were first observed in 1869 by German physicist Johann Hittorf, and were named in 1876 by Eugen Goldstein kathodenstrahlen, or cathode rays.
Thermionic emission	Thermionic emission is the heat-induced flow of charge carriers from a surface or over a pontial-energy barrier. This occurs because the thermal energy given to the carrier overcomes the binding pontial, also known as work function of the metal. The charge carriers can be electrons or ions, and in older lirature are sometimes referred to as 'thermions'.
William Duane	William Duane (February 17, 1872, at Philadelphia - March 7, 1935, in Devon, Pennsylvania) was an American physicist. A coworker of Marie Curie, he developed a method for generating quantities of radon in the laboratory. Biography Studies • 1888-1892 University of Pennsylvania • 1892-1895 Harvard University • 1895 Universities of Göttingen (as a Tyndall Fellow) • 1895-1897 Berlin

----

doctor father: Max Planck

Academic career

- 1898-1907 professor at the University of Colorado
- 1908-1913 at the laboratory of Pierre and Marie Curie in Paris
- 1913-1917 assistant professor of physicsat Harvard University
- 1917-1934 professor of biophysics at Harvard University

research activities

- radioactivity
- X-ray spectroscopy, Duane-Hunt law, relating the minimum wavelength of X-rays to the threshold voltage of the cathode rays that excite them.

Death
Starting in 1925, Duane began suffering a continual decline in health brought on by diabetes.

Work function

In solid-state physics, the work function is the minimum energy (usually measured in electron volts) needed to remove an electron from a solid to a point immediately outside the solid surface (or energy needed to move an electron from the Fermi level into vacuum). Here 'immediately' means that the final electron position is far from the surface on the atomic scale but still close to the solid on the macroscopic scale. The work function is a characteristic property for any solid phase of a substance with a conduction band (whether empty or partly filled).

Carbon nanotube

^{E}Experimental observation; ^{T}Theoretical prediction

The above discussion referred to axial properties of the nanotube, whereas simple geometrical considerations suggest that carbon nanotubes should be much softer in the radial direction than along the tube axis. Indeed, TEM observation of radial elasticity suggested that even the van der Waals forces can deform two adjacent nanotubes. Nanoindentation experiments, performed by several groups on multiwalled carbon nanotubes and tapping/contact mode atomic force microscope measurement performed on single-walled carbon nanotube, indicated Young's modulus of the order of several GPa confirming that Ts are indeed rather soft in the radial direction.

Betatron

A betatron is a cyclotron developed by Donald Kerst at the University of Illinois in 1940 to accelerate electrons, but the concepts ultimately originate from Rolf Widerøe and previous development occurred in Germany through Max Steenbeck in the 1930s. The betatron is essentially a transformer with a torus-shaped vacuum tube as its secondary coil. An alternating current in the primary coils accelerates electrons in the vacuum around a circular path.

Cyclotron

A cyclotron is a compact type of particle accelerator in which charged particles in a static magnetic field are travelling outwards from the center along a spiral path and get accelerated by radio frequency electromagnetic fields.

The cyclotron was invented by Leó Szilárd and first manufactured and patented by Ernest Lawrence, of the University of California, Berkeley who started operating it in 1932 The first European cyclotron was constructed in Leningrad in the physics department of the Radium Institute, headed by Vitali Khlopin.

Flattening

Flattening is a measure of the compression of a circle or sphere along a diameter to form an ellipse or an ellipsoid of revolution (spheroid respectively. Other terms used are ellipticity, or oblateness. The usual notation for flattening is f and its definition in terms of the semi-axes of the resulting ellipse or ellipsoid is

$$\text{flattening} = f = \frac{a-b}{a}.$$

The compression factor is b/a in each case.

Microtron

A microtron is a type of particle accelerator concept originating from the cyclotron in which the accelerating field is not applied through large D-shaped electrodes, but trough a linear accelerator structure. The classic microtron was invented by Vladimir Veksler. The kinetic energy of the particles is increased by a constant amount per field change (one half or a whole revolution).

Annihilation

Annihilation is defined as 'total destruction' or 'complete obliteration' of an object; having its root in the Latin nihil (nothing). A literal translation is 'to make into nothing'.

In physics, the word is used to denote the process that occurs when a subatomic particle collides with its respective antiparticle.

Photon

In physics, a photon is an elementary particle, the quantum of light and all other forms of electromagnetic radiation, and the force carrier for the electromagnetic force. The effects of this force are easily observable at both the microscopic and macroscopic level, because the photon has no rest mass; this allows for interactions at long distances. Like all elementary particles, photons are currently best explained by quantum mechanics and exhibit wave-particle duality, exhibiting properties of both waves and particles.

Storage ring

A storage ring is a type of circular particle accelerator in which a continuous or pulsed particle beam may be kept circulating for a long period of time, up to many hours. Storage of a particular particle depends upon the mass, energy and usually charge of the particle being stored. Most commonly, storage rings are to store electrons, positrons, or protons.

Synchrotron light source

A synchrotron light source is a source of electromagnetic radiation produced by a synchrotron, which is artificially produced for scientific and technical purposes by specialized particle accelerators, typically accelerating electrons. Once the high-energy electron beam has been generated, it is directed into auxiliary components such as bending magnets and insertion devices (undulators or wigglers) in storage rings and free electron lasers. These supply the strong magnetic fields perpendicular to the beam which are needed to convert the high-energy electron energy into light or some other form of EM radiation.

Monitoring	In medicine, monitoring is the evaluation of a disease or condition over time. It can be performed by continuously measuring certain parameters (for example, by continuously measuring vital signs by a bedside monitor), and/or by repeatedly performing medical tests (such as blood glucose monitoring in people with diabetes mellitus). Transmitting data from a monitor to a distant monitoring station is known as telemetry or biotelemetry.
Dual	In category theory, a branch of mathematics, duality is a correspondence between properties of a category C and so-called dual properties of the opposite category C^{op}. Given a statement regarding the category C, by interchanging the source and target of each morphism as well as interchanging the order of composing two morphisms, a corresponding dual statement is obtained regarding the opposite category C^{op}. Duality, as such, is the assertion that truth is invariant under this operation on statements.
Multileaf collimator	A multileaf collimator is a device made up of individual 'leaves' of a high atomic numbered material, usually tungsten, that can move independently in and out of the path of a particle beam in order to block it. MLCs are used on linear accelerators to provide conformal shaping of radiotherapy treatment beams. Specifically, conformal radiotherapy and Intensity Modulated Radiation Therapy (IMRT) can be delivered using MLC's.
Collimator	A collimator is a device that narrows a beam of particles or waves. To 'narrow' can mean either to cause the directions of motion to become more aligned in a specific direction (i.e., collimated or parallel) or to cause the spatial cross section of the beam to become smaller. Optical collimators In optics, a collimator may consist of a curved mirror or lens with some type of light source and/or an image at its focus.

Chapter 14. Particle Accelerators in Medicine

Ionization

Ionization is the process of converting an atom or molecule into an ion by adding or removing charged particles such as electrons or other ions. A substance may dissociate without necessarily producing ions. As an example, the molecules of table sugar dissociate in water (sugar is dissolved) but exist as intact neutral entities.

Ionization chamber

The ionization chamber is the simplest of all gas-filled radiation detectors, and is used for the detection or measurement of ionizing radiation. Conventionally, the term 'ionization chamber' is used exclusively to describe those detectors whh collect ion pairs from gases.

Construction and operation

An ionization chamber is an instrument constructed to measure charge from the number of ions within a medium (whh we will consider to be gaseous, but can also be solid or liquid).

Transmission

A machine consists of a power source and a power transmission system, which provides controlled application of the power. Merriam-Webster defines transmission as: an assembly of parts including the speed-changing gears and the propeller shaft by which the power is transmitted from an engine to a live axle. Often transmission refers simply to the gearbox that uses gears and gear trains to provide speed and torque conversions from a rotating power source to another device.

Laser

A laser is a device that emits light (electromagnetic radiation) through a process of optical amplification based on the stimulated emission of photons. The term 'laser' originated as an acronym for Light Amplification by Stimulated Emission of Radiation. The emitted laser light is notable for its high degree of spatial and temporal coherence, unattainable using other technologies.

Duty cycle

A duty cycle is the time that an entity spends in an active state as a fraction of the total time under consideration. The term is often used pertaining to electrical devices, e.g., switching power supplies. It is also sometimes used pertaining to living systems such as the firing of action potentials by neurons.

Alpha decay

Alpha decay is a type of rioactive decay in which an atomic nucleus emits an alpha particle(two protons and two neutrons) and thereby transforms (or 'decays') into an atom with a mass number 4 less and atomic number 2 less. For example:

238

92U → 234

	90Th + 4 2He which can also be written as: $^{238}U \rightarrow {}^{234}Th + \alpha$, since an alpha particle is the same as a helium-4 nucleus, which has mass number 4 and atomic number 2. It also has a charge +2, but the charge is usually not written in nuclear equations, which describe nuclear reactions without considering the electrons. This convention is not meant to imply that the nuclei necessarily occur in neutral atoms.
GAMMA	GAMMA experiment is a study of: a) Primary cosmic ray energy spectra and elemental composition (abundances of the elements) at energies 10^{15}-10^{18}eV (so called knee energy region) ; b) Galactic diffuse gamma-ray intensity at energies 10^{14}-10^{15}eV ; c) Extensive Air Showers (EAS) at the mountain level by the ground-based EAS array and underground muon scintillation counters; d) Hard jets production at energies ~10^{16}eV by the muon multi-core shower events. GAMMA experiment is deployed on the South side of Mount Aragats in Armenia (Cosmic-ray observatory). The facility consists of a ground-based extensive air shower (EAS) array of 33 surface detection stations and 150 underground muon detectors.
Internal conversion	Internal conversion is a radioactive decay process where an excited nucleus interacts with an electron in one of the lower atom orbitals, causing the electron to be emitted from the atom. Thus, in an internal conversion process, a high-energy electron is emitted from the radioactive atom, but without beta decay taking place. For this reason, the high-speed electrons from internal conversion are not beta partles (β partles), since the latter come from beta decay.

Neutron

The neutron is a subatomic hadron particle which has the symbol n or n^0, no net electric charge and a mass slightly larger than that of a proton. With the exception of hydrogen, nuclei of atoms consist of protons and neutrons, which are therefore collectively referred to as nucleons. The number of protons in a nucleus is the atomic number and defines the type of element the atom forms.

Neutron emission

Neutron emitters to left of lower dashed line
Neutron emission is a type of radioactive decay of atoms containing excess neutrons, in which a neutron is simply ejected from the nucleus. Two examples of isotopes which emit neutrons are helium-5 and beryllium-13. However, the decay of helium-5 is also (by definition) a case of alpha-decay.

Many heavy isotopes, most notably californium-252, emit neutrons among the products of a different radioactive decay process, spontaneous fission.

Proton emission

Proton emission is a ty of radioactive decay in which a proton is ejected from a nucleus.

Proton emission can occur from high-lying excited states in a nucleus following a beta decay, in which case the process is known as beta-delayed proton emission, or can occur from the ground state (or a low-lying isomer) of very proton-rich nuclei, in which case the process is very similar to alpha decay.

For a proton to esca a nucleus, the proton separation energy must be negative - the proton is therefore unbound, and tunnels out of the nucleus in a finite time.

Carl David Anderson

Carl David Anderson was an American physicist. He is best known for his discovery of the positron in 1932, an achievement for which he received the 1936 Nobel Prize in Physics, and of the muon in 1936.

Biography
Anderson was born in New York City, the son of Swedish immigrants.

Avogadro

Avogadro (software) is a molecular editor designed for cross-platform use in computational chemistry, molecular modeling, bioinformatics, materials science, and related are. It is extensible through a plugin architecture.

Features

- Molecular builder/editor for Windows, Linux, and Mac OS X.
- All source code is available under the GNU GPL.
- Translations into Chinese, French, German, Italian, Russian, and Spanish.
- Supports multi-threaded rendering and computation.
- Plugin architecture for developers, including rendering, interactive tools, commands, and Python scripts.
- OpenBabel import of files, input generation for multiple computational chemistry packages, crystallography, and biomolecules.

.

Becquerel

The becquerel is the SI-derived unit of radioactivity. One Bq is defined as the activity of a quantity of radioactive material in which one nucleus decays per second. The Bq unit is therefore equivalent to s^{-1}. The becquerel is named for Henri Becquerel, who shared a Nobel Prize with Pierre and Marie Curie for their work in discovering radioactivity.

William Henry Bragg

Sir William Henry Bragg OM, KBE, PRS (2 July 1862 - 10 March 1942) was a British physicist, chemist, mathematician and active sportsman who uniquely shared a Nobel Prize with his son William Lawrence Bragg - the 1915 Nobel Prize in Physics. im and his son.

Biography Early years
Bragg was born at Westward near Wigton, Cumberland, the son of Robert John Bragg, a merchant marine officer and farmer, and his wife Mary née Wood, a clergyman's daughter.

William Lawrence Bragg

Sir William Lawrence Bragg CH OBE MC FRS (31 March 1890 - 1 July 1971) was an Australian-born British physicist and X-ray crystallographer, discoverer (1912) of the Bragg law of X-ray diffraction, which is basic for the determination of crystal structure. He was joint winner (with his father, Sir William Bragg) of the Nobel Prize for Physics in 1915. He was knighted in 1941. To date, Lawrence Bragg is the youngest Nobel Laureate. He was the director of the Cavendish Laboratory, Cambridge, when the epochal discovery of the structure of DNA was made by James D. Watson and Francis Crick in February 1953.

Chapter 14. Particle Accelerators in Medicine

Curie	The curie is a non-SI unit of radioactivity. It is defined as 1 Ci = 3.7×10^{10} decays per second. Its continued use is discouraged. One curie is roughly the activity of 1 gram of the radium isotope ^{226}Ra, a substance studied by the Curies.
Albert Einstein	Albert Einstein was a German-born theoretical physicist who developed the theory of general relativity, effecting a revolution in physics. For this achievement, Einstein is often regarded as the father of modern physics and one of the most prolific intellects in human history. He received the 1921 Nobel Prize in Physics 'for his services to theoretical physics, and especially for his discovery of the law of the photoelectric effect'.
Enrico Fermi	Enrico Fermi was an Italian-born, naturalized American physicist particularly known for his work on the development of the first nuclear reactor, Chicago Pile-1, and for his contributions to the development of quantum theory, nuclear and particle physics, and statistical mechanics. He was awarded the 1938 Nobel Prize in Physics for his work on induced radioactivity. Fermi is widely regarded as one of the leading scientists of the 20th century, highly accomplished in both theory and experiment.
James Franck	James Franck was a German physicist and Nobel laureate. Biography Franck was born to Jacob Franck and Rebecca Nachum Drucker. Franck completed his Ph.D. in 1906 and received his venia legendi, or Habilitation, for physics in 1911, both at the University of Berlin, where he lectured and taught until 1918, having reached the position of extraordinarius professor.

Hartree

The hartree also known as the Hartree energy, is the atomic unit of energy. It is defined as $2R_\infty hc$, where R_∞ is the Rydberg constant, h is the Planck constant and c is the speed of light. The 2006 CODATA recommended value is E_h = 4.359?743?94 (22) × 10^{-18} J = 27.211?383?86 (68) eV.

Douglas Hartree

Douglas Hartree PhD, FRS (27 March 1897 - 12 February 1958) was an English mathematician and physicist most famous for the development of numerical analysis and its application to the Hartree-Fock equations of atomic physics and the construction of the meccano differential analyser.

Early life

Douglas Hartree was born in Cambridge, United Kingdom. His father, William, was a lecturer in engineering at Cambridge University and his mother, Eva Rayner, was president of the National Council of Women and mayor of the city of Cambridge.

Werner Heisenberg

Werner Heisenberg (5 December 1901 - 1 February 1976) was a German theoretical physicist who made foundational contributions to quantum mechanics and is best known for asserting the uncertainty principle of quantum theory. In addition, he made important contributions to nuclear physics, quantum field theory, and particle physics.

Heisenberg, along with Max Born and Pascual Jordan, set forth the matrix formulation of quantum mechanics in 1925. Heisenberg was awarded the 1932 Nobel Prize in Physics for the creation of quantum mechanics, and its application especially to the discovery of the allotropic forms of hydrogen.

Hertz

The hertz is the SI unit of frequency defined as the number of cycles per second of a periodic phenomenon. One of its most common uses is the description of the sine wave, particularly those used in radio and audio applications. The word 'hertz' is named for Heinrich Rudolf Hertz, who was the first to conclusively prove the existence of electromagnetic waves.

Gustav Ludwig Hertz

Gustav Ludwig Hertz was a German experimental physicist and Nobel Prize winner, and a nephew of Heinrich Rudolf Hertz.

Biography

Hertz was born in Hamburg, the son of Auguste (née Arning) and a lawyer, Dr. Gustav Hertz. He studied at the Georg-August University of Göttingen (1906-1907), the Ludwig Maximilians University of Munich (1907-1908), and the Humboldt University of Berlin (1908-1911).

Max von Laue

Max Theodor Felix von Laue (9 October 1879 - 24 April 1960) was a German physicist who won the Nobel Prize in Physics in 1914 for his discovery of the diffraction of X-rays by crystals. In addition to his scientific endeavors with contributions in optics, crystallography, quantum theory, superconductivity, and the theory of relativity, he had a number of administrative positions which advanced and guided German scientific research and development during four decades. A strong objector to National Socialism, he was instrumental in re-establishing and organizing German science after World War II.

Biography Early years

Laue was born in Pfaffendorf, now part of Koblenz, to Julius Laue and Minna Zerrenner. In 1898, after passing his Abitur in Strassburg, he entered his compulsory year of military service, after which he began his studies in mathematics, physics, and chemistry, in 1899, at the University of Strassburg, the University of Göttingen, and the Ludwig Maximilians University of Munich (LMU). At Göttingen, he was greatly influenced by the physicists Woldemar Voigt and Max Abraham and the mathematician David Hilbert. After only one semester at Munich, he went to the Friedrich-Wilhelms-University of Berlin in 1902. There, he studied under Max Planck, who gave birth to the quantum theory revolution on 14 December 1900, when he delivered his famous paper before the Deutsche Physikalische Gesellschaft. At Berlin, Laue attended lectures by Otto Lummer on heat radiation and interference spectroscopy, the influence of which can be seen in Laue's dissertation on interference phenomena in plane-parallel plates, for which he received his doctorate in 1903. Thereafter, Laue spent 1903 to 1905 at Göttingen. Laue completed his Habilitation in 1906 under Arnold Sommerfeld at LMU.

Career

In 1906, Laue became a Privatdozent in Berlin and an assistant to Planck. He also met Albert Einstein for the first time; they became friends and Laue went on to contribute to the acceptance and development of Einstein's theory of relativity. Laue continued as assistant to Planck until 1909. In Berlin, he worked on the application of entropy to radiation fields and on the thermodynamic significance of the coherence of light waves.

From 1909 to 1912, Laue was a Privatdozent at the Institute for Theoretical Physics, under Arnold Sommerfeld, at LMU. During the 1911 Christmas recess and in January 1912, Paul Peter Ewald was finishing the writing of his doctoral thesis under Sommerfeld. It was on a walk through the Englischer Garten in Munich in January, that Ewald told Laue about his thesis topic. The wavelengths of concern to Ewald were in the visible region of the spectrum and hence much larger than the spacing between the resonators in Ewald's crystal model. Laue seemed distracted and wanted to know what would be the effect if much smaller wavelengths were considered. In June, Sommerfeld reported to the Physikalische Gesellschaft of Göttingen on the successful diffraction of x-rays by Laue, Paul Knipping and Walter Friedrich at LMU, for which Laue would be awarded the Nobel Prize in Physics, in 1914. While at Munich, he wrote the first volume of his book on relativity during the period 1910 to 1911.

In 1912, Laue was called to the University of Zurich as an extraordinarius professor of physics. In 1913, his father was raised to the ranks of hereditary nobility; Laue then became 'Max von Laue'.

Peter Mansfield

Sir Peter Mansfield, FRS is a British physicist who was awarded the 2003 Nobel Prize in Physiology or Medicine, shared with Paul Lauterbur, for discoveries concerning magnetic resonance imaging (MRI). Sir Peter is a professor at the University of Nottingham.

Early Life

Mansfield was born in Lambeth, London on 9 October 1933, to Sidney George (a gas fitter) and Rose Lillian Mansfield.

Peter Kramer

Peter Kramer (physicist) is a German physicist.

Personal life

Kramer studied physics at the University of Münster, the University of Tübingen, the University of Bristol and the University of Marburg. He received his PhD in 1964 in Marburg and in 1968 his Habilitation in Tübingen.

Chapter 14. Particle Accelerators in Medicine

James Clerk Maxwell	James Clerk Maxwell of Glenlair FRS FRSE (13 June 1831 - 5 November 1879) was a Scottish physicist and mathematician. His most prominent achievement was formulating classical electromagnetic theory. This unites all previously unrelated observations, experiments and equations of electricity, magnetism and optics into a consistent theory.
Lise Meitner	Lise Meitner, FRS (7 November 1878 - 27 October 1968) was an Austrian, later Swedish, physicist who worked on radioactivity and nuclear physics. Meitner was part of the team that discovered nuclear fission, an achievement for which her colleague Otto Hahn was awarded the Nobel Prize. Meitner is often mentioned as one of the most glaring examples of women's scientific achievement overlooked by the Nobel committee.
Wolfgang Pauli	Wolfgang Pauli (25 April 1900 - 15 December 1958) was an Austrian theoretical physicist and one of the pioneers of quantum physics. In 1945, after being nominated by Albert Einstein, he received the Nobel Prize in Physics for his 'decisive contribution through his discovery of a new law of Nature, the exclusion principle or Pauli principle,' involving spin theory, underpinning the structure of matter and the whole of chemistry.
Planck	Planck is a space observatory of the European Space Agency (ESA) and designed to observe the anisotropies of the cosmic microwave background (CMB) over the entire sky, at a high sensitivity and angular resolution. Planck was built in the Cannes Mandelieu Space Center by Thales Alenia Space and created as the third Medium-Sized Mission (M3) of the European Space Agency's Horizon 2000 Scientific Programme. The project, initially called COBRAS/SAMBA, is named in honour of the German physicist Max Planck (1858-1947), who won the Nobel Prize for Physics in 1918.
Rayleigh	The rayleigh is a unit of photon flux, used to measure air glow (auroras, for example). It was first proposed in 1956 by D. M. Hunten, Franklin E. Roach, and J. W. Chamberlain. It is named for Robert John Sttt, 4th Baron Rayleigh (1875-1947).
Owen Willans Richardson	Sir Owen Willans Richardson, FRS (26 April 1879 - 15 February 1959) was a British physicist who won the Nobel Prize in Physics in 1928 for his work on thermionic emission, which led to Richardson's Law. Biography Richardson was born in Dewsbury, Yorkshire, England, the only son of Joshua Henry and Charlotte Maria Richardson. He was educated at Batley Grammar School and Trinity College, Cambridge, where he gained First Class Honours in Natural Sciences.

----

1. _________(software) is a molecular editor designed for cross-platform use in computational chemistry, molecular modeling, bioinformatics, materials science, and related are. It is extensible through a plugin architecture.

Features

- Molecular builder/editor for Windows, Linux, and Mac OS X.
- All source code is available under the GNU GPL.
- Translations into Chinese, French, German, Italian, Russian, and Spanish.
- Supports multi-threaded rendering and computation.
- Plugin architecture for developers, including rendering, interactive tools, commands, and Python scripts.
- OpenBabel import of files, input generation for multiple computational chemistry packages, crystallography, and biomolecules.

.

a. Avogadro
b. ESyPred3D
c. Extensible Computational Chemistry Environment
d. ICM-Browser

2. The _________ is the antiparticle or the antimatter counterpart of the electron. The _________ has an electric charge of +1e, a spin of 1/2, and has the same mass as an electron. When a low-energy _________ collides with a low-energy electron, annihilation occurs, resulting in the production of two or more gamma ray photons .

a. Positron
b. Tau neutrino
c. Tribimaximal mixing
d. Trimaximal mixing

3. In physics, a _________ is an elementary particle, the quantum of light and all other forms of electromagnetic radiation, and the force carrier for the electromagnetic force. The effects of this force are easily observable at both the microscopic and macroscopic level, because the _________ has no rest mass; this allows for interactions at long distances. Like all elementary particles, _________s are currently best explained by quantum mechanics and exhibit wave-particle duality, exhibiting properties of both waves and particles.

a. Photon
b. Antiparticle
c. Antimatter comet
d. Antimatter rocket

----

4. Max Theodor Felix von Laue (9 October 1879 - 24 April 1960) was a German physicist who won the Nobel Prize in Physics in 1914 for his discovery of the diffraction of X-rays by crystals. In addition to his scientific endeavors with contributions in optics, crystallography, quantum theory, superconductivity, and the theory of relativity, he had a number of administrative positions which advanced and guided German scientific research and development during four decades. A strong objector to National Socialism, he was instrumental in re-establishing and organizing German science after World War II.

   Biography Early years

   Laue was born in Pfaffendorf, now part of Koblenz, to Julius Laue and Minna Zerrenner. In 1898, after passing his Abitur in Strassburg, he entered his compulsory year of military service, after which he began his studies in mathematics, physics, and chemistry, in 1899, at the University of Strassburg, the University of Göttingen, and the Ludwig Maximilians University of Munich (LMU). At Göttingen, he was greatly influenced by the physicists Woldemar Voigt and Max Abraham and the mathematician David Hilbert. After only one semester at Munich, he went to the Friedrich-Wilhelms-University of Berlin in 1902. There, he studied under Max Planck, who gave birth to the quantum theory revolution on 14 December 1900, when he delivered his famous paper before the Deutsche Physikalische Gesellschaft. At Berlin, Laue attended lectures by Otto Lummer on heat radiation and interference spectroscopy, the influence of which can be seen in Laue's dissertation on interference phenomena in plane-parallel plates, for which he received his doctorate in 1903. Thereafter, Laue spent 1903 to 1905 at Göttingen. Laue completed his Habilitation in 1906 under Arnold Sommerfeld at LMU.

   Career

   In 1906, Laue became a Privatdozent in Berlin and an assistant to Planck. He also met Albert Einstein for the first time; they became friends and Laue went on to contribute to the acceptance and development of Einstein's theory of relativity. Laue continued as assistant to Planck until 1909. In Berlin, he worked on the application of entropy to radiation fields and on the thermodynamic significance of the coherence of light waves.

   From 1909 to 1912, Laue was a Privatdozent at the Institute for Theoretical Physics, under Arnold Sommerfeld, at LMU. During the 1911 Christmas recess and in January 1912, Paul Peter Ewald was finishing the writing of his doctoral thesis under Sommerfeld. It was on a walk through the Englischer Garten in Munich in January, that Ewald told Laue about his thesis topic. The wavelengths of concern to Ewald were in the visible region of the spectrum and hence much larger than the spacing between the resonators in Ewald's crystal model. Laue seemed distracted and wanted to know what would be the effect if much smaller wavelengths were considered. In June, Sommerfeld reported to the Physikalische Gesellschaft of Göttingen on the successful diffraction of x-rays by Laue, Paul Knipping and Walter Friedrich at LMU, for which Laue would be awarded the Nobel Prize in Physics, in 1914. While at Munich, he wrote the first volume of his book on relativity during the period 1910 to 1911.

In 1912, Laue was called to the University of Zurich as an extraordinarius professor of physics. In 1913, his father was raised to the ranks of hereditary nobility; Laue then became '_________'.

a. Max von Laue
b. Leonard Mandel
c. James Clerk Maxwell
d. Albert Abraham Michelson

5. A _________ is a cathode used within nixie tubes, gas discharge lamps, discharge tubes, and some types of vacuum tube which is not electrically heated by the circuit to which it is connected. Such a cathode may be heated to a high temperature in normal circuit operation.

Cold-cathode devices
A cold-cathode vacuum tube does not rely on external heating of an electrode to provide thermionic emission of electrons.

a. Cold cathode
b. Crackle tube
c. Crookes tube
d. Deuterium arc lamp

ANSWER KEY
Chapter 14. Particle Accelerators in Medicine

1. a
2. a
3. a
4. a
5. a

Other Cram101 e-Books and Tests

Lightning Source UK Ltd.
Milton Keynes UK
UKOW02f1325290514

232532UK00001B/9/P

9 781467 274944